# Energy Transitions in Mediterranean Countries

*This book is dedicated to my beloved princess Mariavittoria*
*and my beloved hurricane Enrico*

# Energy Transitions in Mediterranean Countries

## Consumption, Emissions and Security of Supplies

Silvana Bartoletto

*Associate Professor of Economic History, Department of Economic and Legal Studies, University of Naples Parthenope, Italy*

Cheltenham, UK • Northampton, MA, USA

Published by
Edward Elgar Publishing Limited
The Lypiatts
15 Lansdown Road
Cheltenham
Glos GL50 2JA
UK

Edward Elgar Publishing, Inc.
William Pratt House
9 Dewey Court
Northampton
Massachusetts 01060
USA

A catalogue record for this book
is available from the British Library

Library of Congress Control Number: 2020944120

This book is available electronically in the **Elgar**online
Social and Political Science subject collection
http://dx.doi.org/10.4337/9781788977555

ISBN 978 1 78897 754 8 (cased)
ISBN 978 1 78897 755 5 (eBook)

Printed and bound by CPI Group (UK) Ltd, Croydon, CR0 4YY

# Contents

# Introduction

The year 2020 began with high tensions at world level, which brought to the fore the central importance of the Mediterranean region in terms of energy and security of energy supplies. There were tensions between the USA and Iran, the war in Syria, and the recent epilogue of the war in Libya, which saw as protagonists on the one hand Russia, and on the other Erdoğan's Turkey, which decided to send its soldiers to sustain Fayez al-Sarraj against General Khalifa Haftar, a decision not shared by Egypt and other countries, who sought the intervention of the international community. These were events that remind us once again of the strategic importance of the Mediterranean and the need to implement Euro–Mediterranean cooperation. Behind these wars and tensions there are geopolitical interests and issues of primary importance. A major role is played by the desire to acquire control of territories considered strategic because they are both rich in oil and gas and are a place of transit for oil and gas pipelines or ships that transport energy from producing to consumer countries. The Mediterranean region plays a strategic role for energy due to the presence of Libya and Algeria that have large reserves of oil and gas, and because of the discovery of new offshore gas fields in Egypt, Israel and Cyprus. The Mediterranean region is also an important energy transit due to the presence of strategic chokepoints for maritime traffic of oil tankers, such as the Suez Canal and Turkish straits. The role of the Mediterranean region as an energy corridor is reinforced by the presence of pipelines that transport oil and gas from the Middle East, Russia, Azerbaijan and other former Soviet Union States to consumer countries. Moreover, the strategic role of the Mediterranean region is due not only to what already exists, but also to potential future development, which is why the conflicts in Syria and Libya have seen the participation of the superpowers, namely Russia and the USA.

The present book analyses energy consumption, energy production, energy intensity, $CO_2$ emissions and security of energy supply in Mediterranean countries. While previous works focus mainly on European countries, the present book also considers North African and Middle Eastern countries. Our study, conducted in a comparative perspective, shows that diversity and inequality are the main features of the Mediterranean area, which comprises 25 countries that vary greatly in economic terms, development levels and energy consumption. Hence the Mediterranean is rarely considered an economic whole because its various countries are part of three different continents, that is, Europe, Africa and Asia. In contrast to the prevailing literature, we consider the Mediterranean area a single economic system, but at the same time we highlight the differences between disparate areas. The book is completed by an Appendix reporting statistics on energy consumption, energy production, $CO_2$ emissions and other energy indicators calculated considering the Mediterranean area as a whole, including not only European countries, but also North African countries, namely Libya, Algeria, Egypt, Tunisia and Morocco, and Middle Eastern countries, namely Israel, Jordan, Lebanon and Syria.

The red line of the book is the concept of energy transitions: from renewable to non-renewable energy sources; from coal to oil and later natural gas (nuclear for some countries). Analysis of long-term trends of $CO_2$ emissions helps evaluate the effort made by individual countries to reduce the impact of energy consumption on the environment, but at the same time highlights the numerous policy contradictions in support of renewable energies, because fossil fuel consumption continues to grow, especially in emerging economies.

Strictly linked is the analysis of the problem of energy dependency and energy security. As will be shown, the latter is strictly connected to the energy transitions before, during and after World War II. Turmoil and wars in the Middle East and North African countries in more recent years have been triggered by several factors, although the role assumed by energy has been fundamental. However, the current dynamics cannot be understood without examining the problem of energy security from a historical perspective. This book therefore reconstructs not only the most famous oil crises of the 1970s, but also the situation since World War II, when energy security was already a major concern and the exploitation of oil from the Middle East played an important role in the Cold War and the European Recovery Programme. Thus we examine the first oil crisis (1951–1952), strictly linked to the nationalization by the Iranian government of the Anglo–Iranian Oil Company, the Suez Crisis (1956–1957), triggered by the nationalization of the Suez Canal, which is fundamental for the transport of oil from the Middle East to importing countries, the role of Organization of the Petroleum Exporting Countries (OPEC), Libya and the third post-war oil crisis, until the Persian Gulf crises and current energy crises that affect the Mediterranean region. What can be learnt from analysis of the above energy crises is that there are remarkable similarities of current crises with past experience. Recognizing similarities, or simply remembering forgotten history, can help policymakers to better manage present and future crises.

Major similarities can be found in the energy transitions of emerging economies of Middle East and North African countries with past energy transitions of currently more advanced European countries. As will be shown, the study of past energy transitions can be a useful tool for better forecasting the evolution of energy consumption in emerging economies. In more recent years, total and per capita energy consumption have rapidly increased in North Africa and Middle Eastern countries, as have energy intensity, fossil fuel consumption and $CO_2$ emissions. By contrast, energy intensity and $CO_2$ emissions have fallen in the most advanced economies of the countries that are part of the European Union (EU). Analysis of the composition of the energy balance of the various countries further reveals different policies in terms of renewable energy sources. Also through the Gini index, the long-term pattern of inequality in terms of energy consumption and $CO_2$ emissions is estimated. The role of Euro–Mediterranean cooperation for energy and climate change is another major aspect that will be addressed in the book, since the issue of climate change is of key importance for the Mediterranean region, following rapid population growth and urbanization of North African and Middle Eastern countries. The Mediterranean region has great potential for renewable sources, but despite important initiatives such as the 1995 Barcelona Conference and the creation of the Union for the Mediterranean in 2008, the spread of turmoil in the Middle East and North Africa highlights the numerous weaknesses of the Euro–Mediterranean integration process.

## ACKNOWLEDGEMENT

The author is particularly grateful to the University of Naples Parthenope for funding this research project.

# 1. Economy and energy in Mediterranean countries

## 1. ECONOMY AND POPULATION

The Mediterranean countries differ greatly in their institutions, economies and energy consumption. The Mediterranean is an ancient crossroads comprising 25 countries that are part of the Middle East, North Africa, the European Union (EU) and candidate countries to the EU. The 25 Mediterranean countries have many common characteristics but also differ substantially from an economic, cultural, religious and social perspective. Regional inequalities, associated with problems of political instability, wars and serious internal unrest, are fuelling ever-increasing migratory flows from the southern countries, especially from Syria and Libya, but also from Egypt, Morocco, Algeria and Tunisia, towards the richer and safer European countries. Within the Mediterranean region the following six macroareas may be identified: the Latin area, the Adriatic area, the Anatolian–Balkan area, the Middle Eastern area, the Libyan–Egyptian area and the Maghreb. The Latin area includes Portugal, Spain, France, Italy and Malta; the Adriatic includes Slovenia, Croatia, Bosnia and Herzegovina, Serbia, Montenegro, Macedonia and Albania; the Anatolian–Balkan area comprises Greece, Turkey and Cyprus; in the Middle Eastern area are Syria, Lebanon, Israel, Palestine and Jordan; the Libyan–Egyptian area consists of Egypt and Libya; while the Maghreb comprises Tunisia, Algeria and Morocco (Bartoletto, 2016a). The differences are evident both in terms of economic development, and from a demographic standpoint (Capasso, 2018; Zupi, 2017). Indeed, the four main countries of the Latin area, namely France, Italy, Spain and Portugal, account for only 37% of the total population of Mediterranean countries, and yet produce 70% of the total gross domestic product (GDP) and consume 60% of energy.

The total population of the Mediterranean region is about 529 million, amounting to roughly 7% of the world's population.[1] The most populous country is Egypt, with 95.7 million inhabitants, followed by Turkey (78.2 million), France (66.9 million) and Italy (60.6). Taken together, these four countries account for 56.9% of the total Mediterranean population. The remaining 43% is distributed among the other 21 countries. Since the early

1970s until today the population growth has been very rapid, despite the unrest and outright conflict that have occurred in the former Yugoslavia and Syria, Libya and other countries of the Middle East and the Maghreb. The population has risen from 306 million in 1971 to 529 million in 2016, the growth being mainly from North African and Middle Eastern countries, while the more advanced European countries have witnessed a sharp reduction in growth rates (Figure 1.1). For example, the North African population has more than doubled, growing from 74 million to 189 million, and Egypt's population has almost tripled, increasing from 35.9 million in 1971 to 95.7 million in 2016 (Table 1A.2). Over the same period, in Algeria the population has more than doubled, from 15 million to 40.6 million. The same applies to Morocco, where the population has increased from 16.3 million to 35.3 million. The population of North Africa and the Middle East is expected to increase significantly in the coming decades, while the population growth rates of European countries will continue to decline (Zupi, 2017).

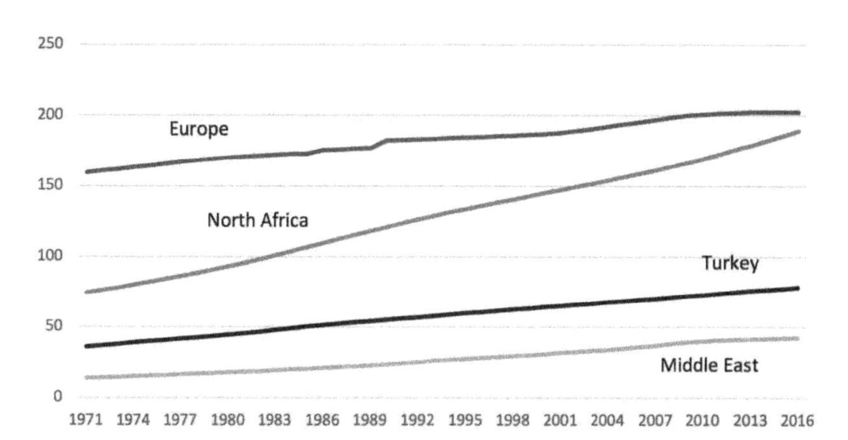


*Figure 1.1* *Population by macroarea in the Mediterranean region, 1971–2016 (millions)*

*Source:* Our calculations based on IEA, Indicators for $CO_2$ Emissions, data extracted on 21 December 2018.

The population factor is very important for security of energy supply: a population increase in countries with oil and gas reserves means a reduction in exportable surplus due to the increase in domestic demand. For example, from being a net exporter of energy, Egypt has become a net importer. By around 2050 the urban population in Mediterranean coastal regions, particularly in coastal cities, is expected to have doubled. Apart from the megapoleis of

Cairo and Istanbul, which are among the largest cities in the world, about 18% of city dwellers live in 85 medium-sized cities, and about half in over 3000 towns with a population of less than 300 000 (Blue Plan, 2012). Typical of the Mediterranean region are the dichotomies between urban and rural areas, the inaccessibility of some areas, and the coastal population density.

The concentration of population in coastal zones is higher in the southern Mediterranean countries. In the countries of North Africa 90% of the inhabitants live in less than 10% of the available area and almost 40% of the population live within 50 km of the coast, with enormous implications in terms of urbanization and vulnerability to possible impacts of climate change. In the Mediterranean region, the growth of urbanization from 1950 to 1995 was very marked (Blue Plan, 2001). During that period, the countries showing the highest urbanization rates were Libya and Turkey. The spectacular growth of cities in Libya has been linked to a profound change in the economy brought about by income from oil. Libya is almost all desert or semi-desert, with a small population that the central government has sought to settle in coastal cities. Yet the growth of such settlements holds out the prospect of worsening environmental problems due to higher energy consumption and $CO_2$ emissions, excessive land consumption, aquifer pollution, inefficient waste management and the cumulative effect of all these factors on the environment and human health.

Under the combined effect of demographic pressure and economic growth, the region is witnessing a sharp increase in energy demand, especially for electricity. Demand is expected to rise four to five times faster in the Southern and Eastern Mediterranean Countries (SEMCs)[2] than in the countries on the northern shore (Blue Plan, 2012). Energy is one of the main factors which leads to an increase in urban ecological footprint, and the growth of the number and size of cities is one of the main factors behind rising energy consumption. In the SEMCs the fossil fuel dependency of countries and hence of cities has reached 90%. The growing consumption of oil products is closely linked to development in the transport sector. The increasing production of electricity from fossil fuels, which is the response to growing demand for electricity driven mainly by the intensive use of electrical equipment, is set to continue especially in North Africa and Middle Eastern countries over the coming years.

There are large divides within the Mediterranean region not only with regard to population dynamics, but also from an economic perspective: the more affluent countries that are part of the EU, namely Malta, France and Italy, followed by Slovenia, Spain, Portugal and Greece, contrast with their developing counterparts on the eastern shores and especially those on the southern shores, namely Algeria, Egypt, Libya and Morocco (Figure 1.2). In Syria, in 2010, per capita GDP was about $6300, but as a result of the ongoing war, the economy has been devastated, and by 2016 per capita GDP had plunged to $1842 (all amounts in US dollars). Also in Libya, GDP per capita has further decreased, because of the unrest and political and economic instability that the country is still experiencing. In 2010, before the end of the Gaddafi regime, per capita GDP was about $28 887, while in 2016 it had fallen to $7143 (Table 1A.1).[3]

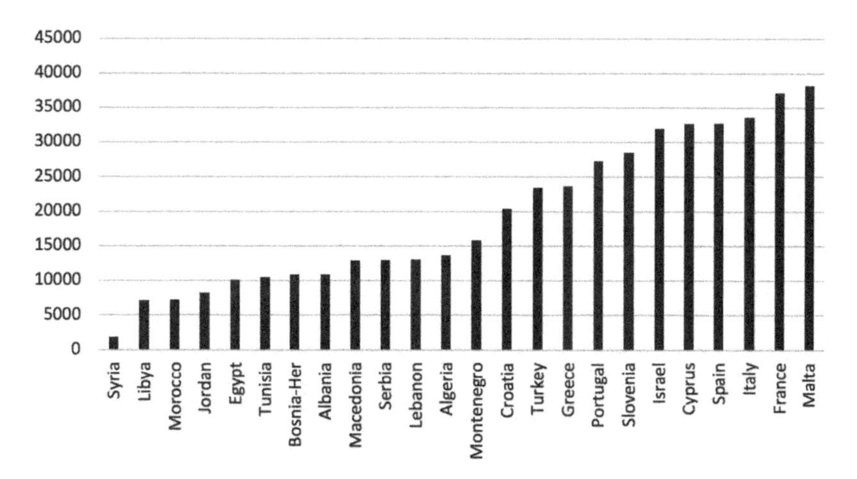


*Figure 1.2* *Per capita GDP in Mediterranean countries in 2016 (2010 USD)*

*Source:* Our calculations based on IEA, Indicators for $CO_2$ Emissions, data extracted on 21 December 2018.

In Southern Mediterranean countries, the share of the population below the minimum poverty line ($1 per day) is low, but increases drastically if the threshold of $2 per day is taken (Ferragina and Nunziante, 2018). A large part of the population is therefore subject to the slightest negative shock, for example when food and fuel prices increase. Thus the per capita GDP decreases progressively as one moves geographically towards the Balkans, the Middle East and North Africa. Of interest is the case of Slovenia, which recovered immediately after independence, reaching a significant level of development. The economic crisis that began in 2008 had very negative effects on Greece and Portugal. Italy too has been struggling with low economic growth, high unemployment and high public debt. It may be stated that due to the recession and slow economic growth of the previous ten years, Italy's GDP has returned in real terms to the same levels as the late 1990s. In the Maghreb and in the Middle East, development rates declined, but the areas were less affected by the 2008 crisis. Starting from the end of the Gaddafi regime in 2011, Libya's real income suffered a dramatic contraction as a result of the civil war, which halted oil production and exports on which the Libyan economy depended. Indeed, according to International Monetary Fund (IMF) estimates, the hydrocarbon sector provided the Libyan economy with more than 90% of government revenues and export earnings (IMF, 2013).

Algeria was able to benefit from the rise in oil prices and natural gas, accounting for around 70% of government revenues. Despite the long civil wars, the problem of terrorism and Islamic fundamentalism, Algeria has experienced a major economic recovery, due especially to the high revenues from oil and natural gas exports, even though there has been a slowdown in growth rates in recent years.

In Egypt, after a period of intense growth, there was a slowdown due to the 2011 revolution, which led to the end of the Mubarak regime. Income from tourism has declined together with capital inflows, while the unemployment rate has increased.

Jordan experienced a period of intense development, with a GDP growth rate that reached an annual average of 6.7% in the period 2000–2006. Since 2010, growth rates have decreased significantly due to the international economic situation. Jordan has been hit by a series of severe shocks, such as conflicts in neighbouring Syria and Iraq, which caused considerable economic and social pressure (IMF, 2015). In the Mediterranean region, Jordan is considered one of the most open economies, and is heavily reliant on revenues from tourism, remittances from Jordanians' overseas and foreign investment, as well as being very dependent on foreign countries to meet its energy needs.

Since the recession of 2009 caused by the European crisis and the fall in domestic demand, Turkey has largely recovered thanks to far-reaching structural reforms implemented in the tax system and in the labour market. It is also one of the Mediterranean countries where population growth has been very intense, rising from 36 million in 1971 to 78 million in 2016.

There are also far-reaching divides in economic structures. The process of economic development generally involves a change in the weight of individual sectors, with a reduction in the agricultural sector and an increase in the industrial and service sectors. In the Mediterranean area, alongside the most advanced economies with a prevalence of industry and services, some countries show a delay in development, where agriculture continues to play a major role both in terms of workforce share and added value. By contrast, in the most advanced economies such as France, Italy and Spain, the service sector predominates both in terms of added value and numbers of employees. Indeed, in the above countries, agriculture accounts for merely 3–4% of the workforce, industry around 20–25%, while services employ more than 60%. The economic structure is completely different in North Africa where agriculture still accounts for a high share of the workforce, amounting to about 28% in Egypt, and as much as 39% in Morocco (Ansani and Daniele, 2014). In contrast, in oil- and natural gas-rich countries in North Africa, the main production activity is the exploitation of such natural resources.

In analysing inequality in Mediterranean countries it is important to distinguish between different periods. Ansani and Daniele (2014) show that from 1950 to 2000, inequality grew, arguing that, rather than economic convergence, Mediterranean countries experienced divergence. A period of decline occurred in the second half of the 1970s when the advanced economies of the North went through a period of crises due to the increase in oil prices, whereas the producing and exporting SEMCs such as Algeria, Libya, Egypt and Syria benefited from an increase in revenues thanks to the growth in oil revenues. However, from the early 1980s onwards, the gap increased once again and peaked at the beginning of the 1990s, remaining stable at these high levels throughout the decade. Contrary to expectations, a convergence process has taken place since 2001, thanks to the reduction in the gaps that returned to the levels of the early 1980s. The great crisis that began in 2007 hit European countries hard, especially Greece, Portugal and Italy, where there has been a contraction in GDP per capita, a reduction in production and an increase in unemployment, not only due to the crisis but also to restrictive fiscal policies. The sovereign debt crisis in Greece, Portugal, Spain and Italy has led to the adoption of restrictive fiscal policies which, added to the effects of the crisis, have produced recessionary effects on the economy, whose consequences have reflected in the levels of energy consumption. Thus from 2001 to 2011, differences between areas were reduced, not only because the growth rates of the countries lagging behind were higher but also because in the advanced countries there was a reduction in growth rates. Thus, in the Mediterranean region, convergence was achieved because the stronger countries declined and the weaker countries advanced (Bartoletto and Malanima, 2014).

However, because of the conflicts and serious unrest that began in 2011 with the so-called Arab Spring, in the countries of the Middle East and North Africa there has been a slowdown in growth rates. The still ongoing war in Syria and Libya made the picture even more complex. When Russia intervened in the war in Syria in September 2015, its goal was to ensure the survival of the Assad regime and to reduce US influence in Syria. After Russia's agreement with Turkey on 23 October 2019, Russia's "victory" in Syria was somehow sanctioned and an important role also for Turkey was envisaged.[4] At the time of writing, the alliance between Putin and Erdoğan, consolidated after the attempted coup in Turkey in 2016, is also playing an important role in resolving the Libyan issue. The year 2020 started with growing concern at global level, because on 3 January a US raid on Baghdad airport in Iraq caused the death of Iranian general Qasem Soleimani, a key figure in the Ayatollah regime. On the night of 8 January a number of missiles hit the bases of Ain Al-Asad, one of the largest US installations in Iraq, and Erbil, the capital of Iraqi Kurdistan.

Shortly thereafter, Tehran TV announced the start of the "Soleimani martyr" operation in response to the killing of the general. Iran carried out a massive offensive. At least 13 ballistic missiles left Tehran territory for US bases.

In Israel, the maximum alert was triggered and the price of oil on international markets immediately began to rise. To complicate the global picture there is the war in Libya, whose recent developments seem to be leading towards a division of its territories between Russia and Turkey. Indeed, on 10 December 2019, Turkish President Erdoğan announced his intention to militarily support the Libyan national agreement (GNA) government led by Fayez al-Sarraj against the military offensive of General Khalifa Haftar. And less than a month after its announcement, on 2 January 2020, the Turkish parliament authorized the sending of soldiers to Libya, approving the resolution with 325 votes in favour and 184 against.

## 2. DYNAMICS OF TOTAL AND PER CAPITA ENERGY CONSUMPTION

Increasing population and economic development has led primary energy consumption in the Mediterranean region to more than double, rising from about 393 million tons of oil equivalent (Mtoe) in 1971 to about 978 Mtoe in 2016 (Figure 1.3).

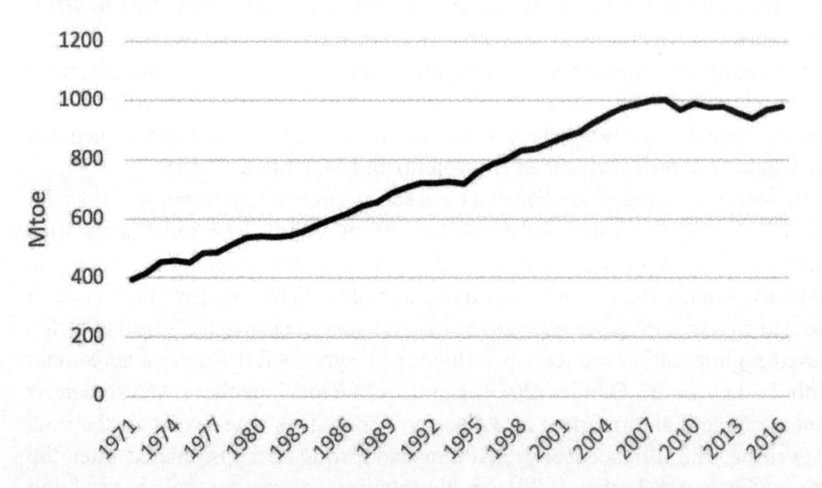


*Figure 1.3* *Total energy consumption in Mediterranean countries, 1971–2016 (Mtoe)*

*Source:* Our calculations based on IEA, World Indicators, data extracted on 15 July 2019.

The growth of energy consumption in the Mediterranean region is in line with global energy dynamics, since world energy consumption more than doubled during the same period of time, growing from 5523 in 1971 to 13 761 Mtoe in 2016.[5] However, the pace of growth has greatly diminished in recent years, and there is also a marked difference between the Organisation for Economic Co-operation and Development (OECD) and non-OECD countries: while in OECD countries there has been a reduction in the energy consumption rate, non-OECD countries have experienced a significant increase. Currently, OECD countries consume 38% of the world total, compared to 61% in the early 1970s. Much of the growth in energy consumption in the non-OECD group is due to China, which recently overtook the USA in terms of total energy consumption. There are several reasons that led to a reduction in energy consumption in OECD countries: first of all, the serious economic crises that hit the most advanced economies hard, causing a reduction in production and hence in energy consumption; but also other factors, such as improving energy efficiency and changes in the composition of the energy basket. A decisive contribution was made by the USA, where in recent years there has been a decrease in primary energy consumption thanks to the gradual replacement of coal with natural gas in electricity generation.[6] However, the USA maintains a dominant position within the OECD countries, on its own accounting for 41% of the regional total, slightly less than in 1971.

The situation appears different if we consider per capita rather than total consumption: while the USA accounts for only 4% of the world's population, contrasting with China's 19%, per capita consumption in the USA is about three times higher than in China. In 2016 per capita consumption in the USA was 6.69 toe, compared with 2.14 in the People's Republic of China.[7] In general, the OECD countries are those that have the highest per capita consumption for different reasons: the very high level of electrification, which is close to 100%, the larger number of cars per household, a very high GDP and a strong share of industry and services out of total GDP. Instead, the level of energy intensity (energy consumption by GDP) is below the world average. This is due to the high GDP, a growing level of energy efficiency, especially in transformation processes, from high efficiency in final consumption (machines that consume less fuel, better insulation of homes) and from relocation of high-energy intensity industries.

The divide between OECD and non-OECD countries is also reflected in the Mediterranean region. Indeed, North African and Middle Eastern countries have been affected by a sharp increase in energy consumption, triggered by consumer subsidies, which have caused overconsumption. The effects are levels of energy intensity higher than the worldwide average, and hence below-average energy efficiency (Jalilvand and Westphal, 2018), aside from higher $CO_2$ emissions. In the countries of North Africa, total energy consumption rose more than tenfold, from about 18 Mtoe in 1971 to 185 in 2016 (Figure 1.4). While in 1971 North Africa represented only 4% of total consumption of the Mediterranean area, it currently represents 19% (Figure 1.5). On the other hand, while EU member states accounted for 81% of total consumption in 1971, by 2016 their share had fallen to 59%. If we consider the Latin area in the strict sense (Italy, France, Malta, Portugal and Spain), the reduction is even greater, from 80% to 55%.

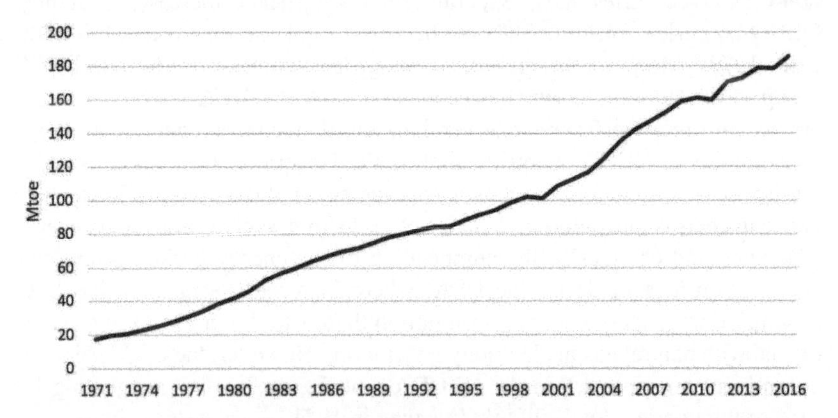


*Figure 1.4* *Total energy consumption in North Africa, 1971–2016 (Mtoe)*

*Source:* Our calculations based on IEA, Indicators for $CO_2$ Emissions, data extracted on 21 December 2018.

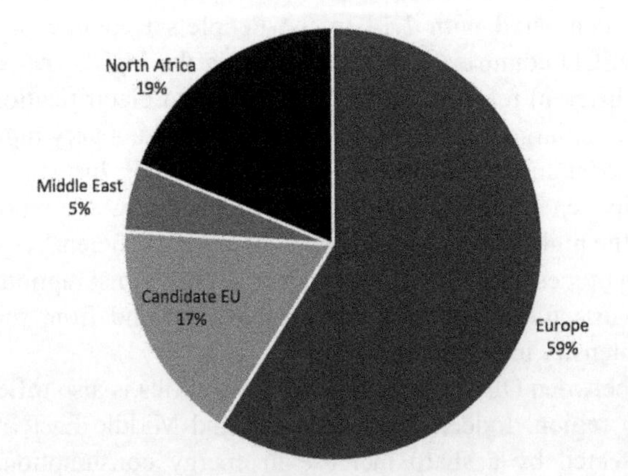


*Figure 1.5* *Total energy consumption in Mediterranean countries by macroregion, 2016 (%)*

*Source:* Our calculations based on IEA, Indicators for $CO_2$ Emissions, data extracted on 21 December 2018.

Different growth rates over time have led to a reduction between different macroareas of the Mediterranean region, since the non-OECD countries have recorded higher growth rates than the most advanced countries of the Latin area (Bartoletto, 2016a).

With regard to per capita energy consumption, the Mediterranean countries are very dissimilar. In the first ten positions of countries with the highest per capita energy consumption we find France, Slovenia, Israel, Spain, Cyprus, Italy, Libya, Serbia, Portugal and Greece (Figure 1.6).

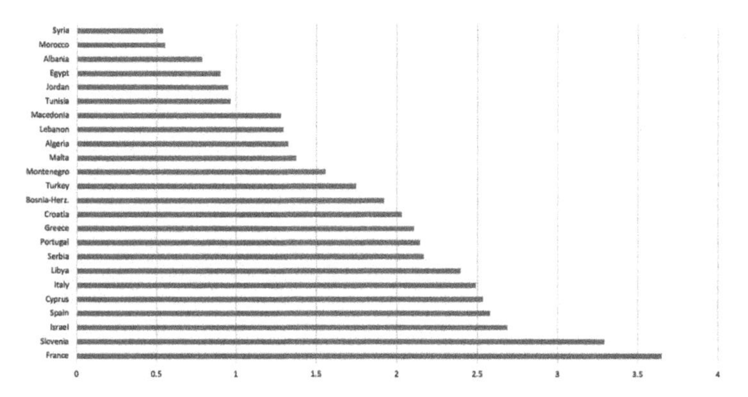


*Figure 1.6* *Per capita energy consumption in 2016 (toe)*

*Source:* Our calculations based on IEA, World Indicators, data extracted on 2 January 2019.

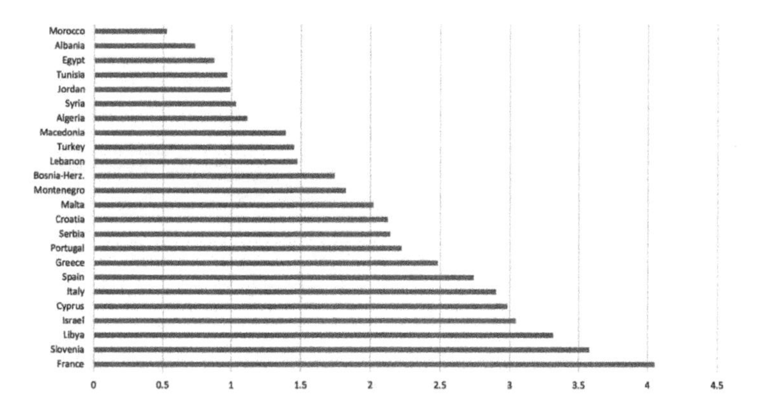


*Figure 1.7* *Per capita energy consumption in 2010 (toe)*

*Source:* Our calculations based on IEA, World Indicators, data extracted on 2 January 2019.

The gaps in per capita consumption levels are considerable. In France per capita consumption is 10 times higher than in Morocco, which is the country with the lowest per capita consumption, along with Syria (Figure 1.6). Not only is there a divergence between the countries of the North and those of the South, but even within the individual areas the differences are considerable. In the North Africa region, Libya had the highest levels of per capita energy consumption throughout the period analysed (Figure 1.8). The growth was very rapid after the discovery of huge oil reserves at the beginning of the Gaddafi regime in 1969, which made Libya the richest country in the North African region in terms of GDP per capita. Following the spread of the Arab Spring and the end of the Gaddafi regime in 2011, per capita energy consumption fell drastically, together with GDP per capita. Since the collapse of oil and gas production, which represented almost the only source of revenue, the ongoing serious internal unrest has not allowed the country to kickstart its recovery. Yet despite the substantial decrease in energy consumption, Libya is still the country with the highest levels of per capita energy consumption among its North African neighbours. Indeed, from Figure 1.8, there is an evidently large divergence with Algeria, Egypt, Morocco and Tunisia, whose per capita energy consumption lagged far behind. Algeria ranks second in terms of such energy consumption, followed by Tunisia, Egypt and, last of all, Morocco.

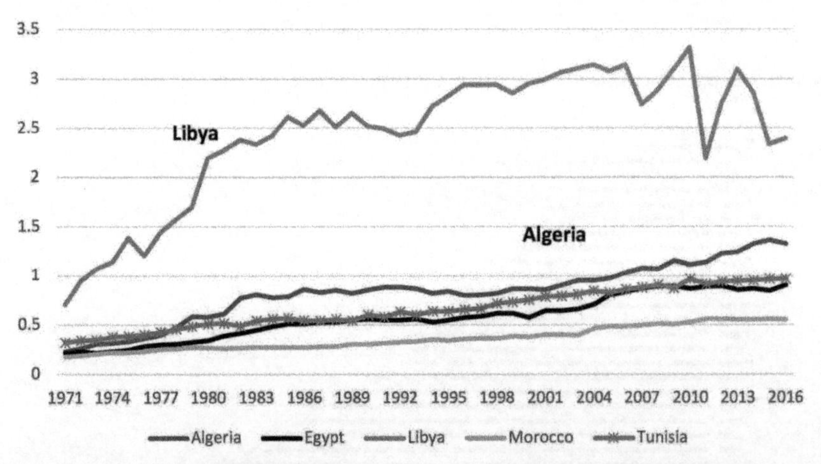


*Figure 1.8* *Per capita energy consumption in North African countries, 1971–2016 (toe)*

*Source:* Our calculations based on IEA, World Energy Balances, data extracted on 21 December 2018.

The same applies to Syria, where the economic picture has radically changed in recent years, following the outbreak of war. It is notable that before the war, per capita energy consumption in Syria was higher than that of Egypt, Tunisia or Jordan (Figure 1.7).

In order to ascertain whether the gap between the different countries is narrowing, we calculate the Gini index on per capita energy consumption, considering the period 1971–2016 (Figure 1.9). The Gini index is the most common statistical index of diversity or inequality to measure the dispersion of a distribution in ecology and social sciences. It is also widely used in econometrics as a standard measure of inequality in income and wealth. It ranges from 0, which represents perfect equality (all measurements are equal), to 1, which represents the maximum degree of inequality (Gini, 1921; Bellù and Liberati, 2006). Inspection of Figure 1.9 shows that the Gini index, calculated on per capita energy consumption of each Mediterranean country, has decreased over time. The variation range fell from 0.41 in 1971 to 0.28 in 2016. Estimation of the index confirms that the reduction of the gap concerns not only total energy consumption, but also per capita energy consumption. It may thus be argued that the growth in total consumption in North African and Middle Eastern countries is not only due to population growth, but also to an increase in per capita consumption.

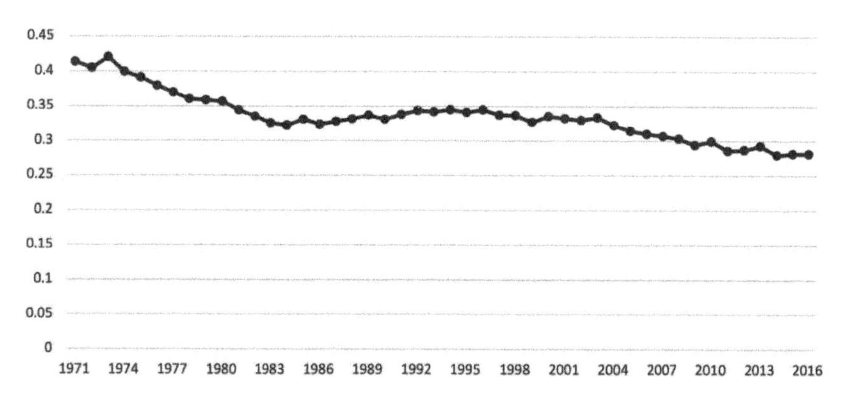


*Figure 1.9* *Gini index on per capita energy consumption, 1971–2016*

*Source:* Our estimates based on IEA, World Energy Balances, data extracted on 21 December 2018.

The above results may be explained by considering changes in the energy basket over the years. Since the energy transition has led to a reduction in consumption of traditional sources and an increase in modern sources, in the countries on the southern and eastern shores of the Mediterranean, which are less industrialized than those on the northern shore, this process of substitution occurred later and, in recent decades, there has actually been a convergence in the consumption of modern energy sources.

On calculating the growth rates of the Gini index, it emerges that, with reference to the per capita energy consumption, the narrowing of the gaps between the Mediterranean countries was considerable between 1971 and 1981. During this decade, the gap decreased by about 16%. There are several reasons that could account for this trend. On the one hand, the richest countries in Mediterranean Europe were experiencing a serious economic and financial crisis due to the oil crisis and the serious monetary disruption created by the end of the Bretton Woods system. On the other, the North African and Middle Eastern countries were thriving on the rich proceeds from oil exports and consuming increasing amounts of fossil fuels. In the following decade 1981–1991, the growth rate fell back drastically, with the gaps in per capita energy consumption continuing to decrease, but at an estimated average rate of 1.7%. During this phase, the negative consequences of previous conflicts were felt, such as the war between Israel, and Egypt and Syria. The difficult situation led to a reduction in oil exports and related income, which all had repercussions on narrowing the gap between the richest and poorest countries in the Mediterranean. In the following decade 1991–2001 the reduction in the gaps in terms of per capita consumption continued, but at a slightly lower annual average rate of 1.6%. This was due above all to the recovery of energy consumption in the countries belonging to the Latin area. However, the dynamics of per capita consumption are also affected by the severe economic crisis in the countries that form part of the Adriatic area. Albania, subjected from 1928 to the Italian fascist regime, regained its independence in 1944, being subsequently governed for decades by a strict communist regime. In 1990, the severe internal economic crisis fuelled a strong popular protest, which led to the end of the previous regime, also due to the collapse of the Soviet Union. A process of transition towards parliamentary democracy was therefore initiated which was far from smooth, fuelling in the process mass migratory phenomena, and resulting in a political and social crisis that led to an armed uprising. The intervention of a United Nations (UN) military mission in 1997 allowed the country to stabilize and political elections to be held. Since then in Albania there have been successive centre-right and centre-left governments. Albania is currently one of the candidate countries to join the EU.

However, also the break-up of the Yugoslav Socialist Republic, which was established in 1945, contributed to slowing the narrowing of the gap. Slovenia, formerly part of the Yugoslav Socialist Republic, declared itself independent in June 1991, and obtained recognition from the international community in January 1992. In 2004 it joined the EU, and in 2007 it adopted the euro. The break-up of the former Yugoslavia left many unresolved issues among the republics that formed part of it. Croatia, also part of the former Yugoslavia, declared itself independent in June 1991, and was recognized by the international community in October of the same year. This led to a bitter conflict with the Serbian minority calling for independence. From 1992 to 1995 Croatia participated in the conflict in Bosnia and Herzegovina. The latter declared its independence on 9 January 1992. The same year a bloody civil war began between Croatian, Bosnian-Muslim and Serbian nationals, only ending after the intervention of the UN and NATO. Macedonia, also a federated republic within Yugoslavia, declared its independence on 15 September 1991, though only achieving recognition by the international community in 1993, under the name of the Former Yugoslav Republic of Macedonia (FYROM).

In conclusion, the recovery in energy consumption of European countries and the crisis in the countries belonging to the Adriatic area impacted on the dynamics of the gap between the most advanced and the least advanced countries in the decade from 1991 to 2001. The gap narrowed considerably during the following decade (2001–2010), the average annual reduction being estimated at 13%. Energy consumption is increasing in the economies of North Africa and the Middle East, and the process of replacing conventional sources of energy with so-called modern sources continues, especially in Morocco and Tunisia. At the same time, however, the most advanced economies are in recession due to the 2007 crisis, which led to a reduction in energy consumption.

Our estimates of the Gini index conclude with the analysis of the last period for which we have data available, that is, the years from 2011 to 2016. The reduction in the gap continues, but the relative rate falls to 1.6%. The so-called Arab Spring, which broke out in 2011, affected energy consumption. Started with the aim of opposing dictatorships and spreading democracy, it ultimately proved to be a failure, since it led to conflict, internal turmoil and the disruption of oil and gas production.

## 3. ENERGY CONSUMPTIONS BY FUEL

The composition of the world energy balance has considerably changed over time. Oil remains the main energy source in the world, although its share has declined from 44% in 1971 to 32% in 2016. The reduced share of oil as an energy source is to be explained by the increase in other energy sources: first natural gas and then nuclear and coal. Although conventional oil still plays a prominent role, the pace of discovery has plunged in the last decade. Thus there is an over-reliance on existing reserves to satisfy future demand.

It is believed that the rapid entry into operation of shale oil deposits in the USA, combined with technological progress, will make it possible to meet future demand. However, according to some estimates, shale oil will continue to represent a mere 10% of the global oil market. Conventional oil will therefore continue to play a key role.

With regard to the pace of conventional oil discoveries, between 2000 and 2014 an average of 10 billion barrels a year were discovered. However, in the last three years this amount has fallen to just 3 billion due to the reduction in exploration spending (Chauhan and Sen, 2018). Producers justify the fall in the reserves to production ratio by claiming that the industry no longer needs a large reserve margin, which would constitute an inefficient use of capital. The exhaustion of oil wells and their decline are two fundamental aspects of the debate on the future of the oil supply. Rates of decline and exhaustion are interconnected but are not the same thing. Rates of decline refer to the annual reduction in the rate of production from a field after it has reached its peak production. The rate of exhaustion instead refers to the rate at which oil is produced by a reservoir, expressed as an estimate of the fraction of the extractable quantity (estimated ultimate recovery–EUR) or of the remaining reserves. Generally an oil field begins its declining phase after 40% of its reserves have been extracted. Future increasing demand cannot be satisfied by relying only on existing reserves. About 70% of non-OPEC reserves have been exploited and it is expected that in the following decades the exploitation rate will be higher than that deemed sustainable. The decline in oil consumption at world level has been partially offset by natural gas, whose consumption increased from 16% in 1971 to 22% in 2016 (Figure 1.10). In the USA, gas electricity generation grew from 18% in 2005 to 31% in 2017. Also in Europe, the role of gas in electricity generation has significantly increased over time. Unlike the oil market, the gas market will continue to be oversupplied for the next few years. However, forecasts are changing rapidly due to the growth in gas consumption. Thus the coming years are likely to see an absorption of the overcapacity of past years. The growth in demand will depend on the emerging countries of Asia and above all China. The Middle East is also highly likely to play a leading role.

The main drivers of the increase are industry and electricity generation. In recent years, the market has been revolutionized by non-conventional gas, namely shale gas, the natural gas stored in porous rocks, which it has become possible to extract thanks to the progress of technology, especially fracking – with which liquids are injected into rocks with high-pressure jets – and horizontal drilling. Shale gas represents more than half of the new reserves ascertained today and in the USA it has increased the resources available threefold. In Europe the most significant deposits are believed to be in Poland and France.[8] Shale gas and also shale oil, both of which are hydrocarbons contained in shale rocks, have been the main protagonists of the energy "revolution" in recent years. In addition to purely energy-based importance, the exploitation of these unconventional sources has had far-reaching political and economic implications worldwide. The first to realize the potential of this resource was the USA which, once the appropriate technology was developed, began to extract it. In just a few years the production of such unconventional hydrocarbons has allowed the USA not only energy independence but also the possibility of exporting surplus gas and oil abroad, with “side-effects” on all the world’s energy markets. This huge quantity placed on the market has produced, at the global level, since mid-2014, a fall in the price of crude oil. Only recently have OPEC and non-OPEC countries (especially Russia) found measures to counteract this oversupply with an agreement to cut production.

Although coal is the most polluting fossil fuel, it continues to be the second most important energy source globally, accounting for a 27% share of total energy sources in 2016.[9] Nonetheless, in the USA coal-based electricity generation decreased from 40% in 2014 to 31% in 2017, to the advantage of gas. The important position of coal consumption at world level stems largely from the increased demand from China, which has continued to invest in power plants fuelled by coal for electricity production. China currently produces about 45% of the world’s total, followed, at a considerable distance, by India, with 9.7%, and the USA, with a share of 9.3%. Yet there is also a resurgence of coal consumption in other countries such as Germany, which in recent years has implemented the production of lignite, the most polluting coal. Germany is currently ranked eighth in the world for coal production.[10]

Nuclear energy represents about 5% of total demand, four percentage points more than in 1971, although several governments have announced their aim to close down nuclear power plants following the Fukushima disaster in 2011. The main producer is the USA, which accounts for 840 TWh of nuclear electricity, amounting to 32% of world nuclear electricity production.[11] France is second in the world ranking with a production of 403 TWh and a share of 17.2 % of the world total. The People’s Republic of China comes third, with a production of 213 TWh of nuclear electricity, corresponding to 8% of the world total. In fourth position is the Russian Federation, with a production of 197 TWh and a share of about 7%. Germany continues to be in the top ten of producing countries, although the German government declared soon after the Fukushima disaster its objective to reduce nuclear production. It now produces 85 TWh, corresponding to about 3.3% of world total.

Despite the extensive debate on climate change and the negative consequences for the environment of fossil fuels and nuclear consumption, renewable energy at world level represents only 14% of total consumption, only one percentage point more than in 1971 (see Figure 1.10). This represents a disconcerting contradiction between what the governments of the various countries announce and what is then concretely done in terms of investments in renewables (Clô, 2017), even though in some countries major efforts have been made to achieve greater protection of the environment.

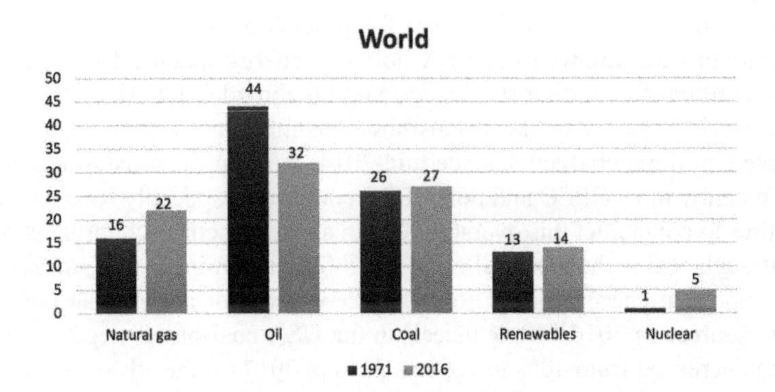


*Figure 1.10* *World energy consumption by fuel, 1971 and 2016 (%)*

*Source:* IEA, World Energy Balances, data extracted on 20 November 2018.

The energy balance in the Mediterranean region is somewhat different: coal plays a more limited role, and there has been significant progress in renewable energy, even though there are major differences between and within different areas (Figure 1.11). Indeed, the rapid growth of fossil fuel consumption in the developing countries of North Africa and the Middle East has made the Mediterranean's share of renewable energy lower than its global counterpart.

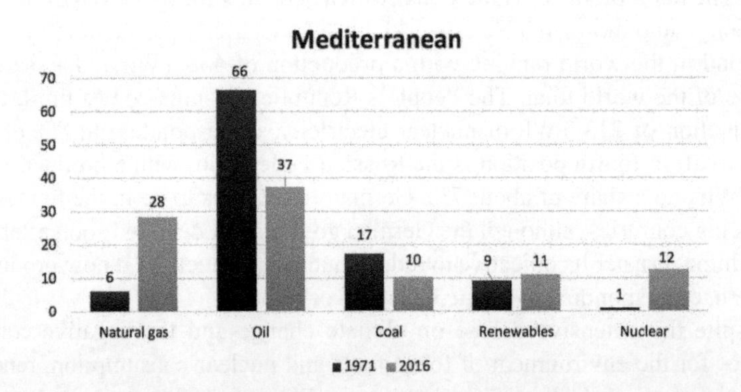


*Figure 1.11* *Energy consumption by fuel in the Mediterranean region, 1971 and 2016 (%)*

*Source:* Our calculations on IEA, World Energy Balances, data extracted on 20 November 2018.

The energy balance of the Mediterranean region is also dominated by fossil fuels, which make up 76% of total primary energy consumption (Figure 1.11). Oil remains the main energy source, with a 37% share of the total, even though over the years its consumption has decreased in both absolute and relative terms. The share of natural gas increased from 6% in 1971 to 28% in 2016, while that of coal decreased from 17% to 10% (Figure 1.11). Analysing oil consumption from 1971 to today, we note that the peak was reached in 2006 (Figure 1.12), with a consumption of 416 594 kilo tons of oil equivalent (ktoe). Its consumption then decreased, falling to 363 949 ktoe in 2016. Natural gas consumption has increased considerably. In 1971, its consumption was only 24 193 ktoe, but following the oil crises of 1973 and 1979, its consumption grew very rapidly in the Mediterranean region, peaking at 277 482 ktoe in 2010. Subsequently, as a result of the 2007 crises, natural gas consumption started to decrease, but after the low of 263 861 ktoe in 2014, it rose to 281 950 in 2016 (Figure 1.12).

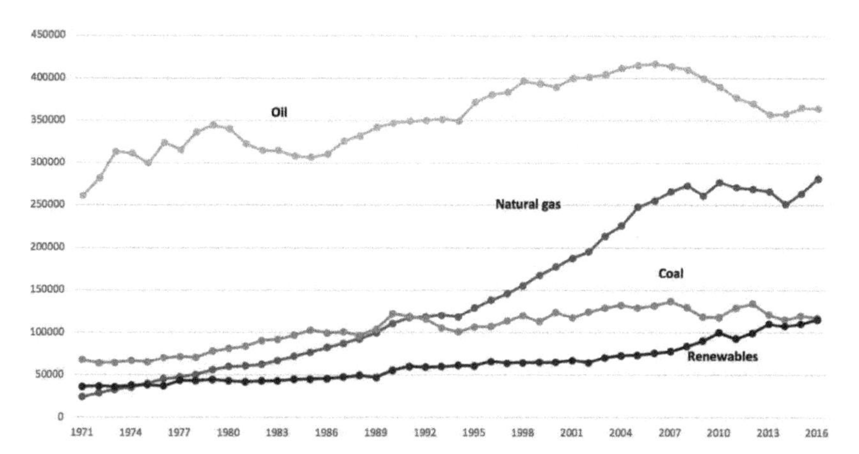


*Figure 1.12* *Consumption of renewables and fossil fuels in Mediterranean countries, 1971–2016 (ktoe)*

*Source:* IEA, World Energy Balances, data extracted on 20 November 2018.

Coal instead satisfies only 10% of total demand, a value well below the world average (Figure 1.11). Coal consumption is mainly concentrated in Turkey, Israel, Morocco and former Yugoslavia, which have the highest proportion of coal consumption with respect to the other countries of the Mediterranean region.

The growth of nuclear electricity was remarkable in the Mediterranean region after the oil crises. Indeed, nuclear rose from 1% of total energy consumption to 12%, and now represents the third energy source in order of importance. Yet about 86% of nuclear consumption is concentrated in France, which represents at global level the country with the highest percentage of nuclear-generated electricity. Due to France alone, the percentage of nuclear consumption in the Mediterranean region is well above the world percentage. As a result of the massive use of nuclear energy, in France the consumption of oil has been significantly reduced, not only in percentage terms compared to other energy sources, but also in absolute terms. In 1992 nuclear consumption overtook oil consumption and the gap has continued to widen ever since. The use of nuclear power has allowed France not only to drastically reduce its energy dependence on fossil fuels, but also to reduce its electricity costs: in France, the cost of one kWh of electricity is as much as 40% lower than in Italy, which has no nuclear option in its energy basket.

Within the Mediterranean region, apart from France, only two other countries have nuclear power plants: Spain, which has eight nuclear power plants, and Slovenia, which has only one. In 1964, the leading Spanish electricity company began an ambitious project to construct nuclear power plants. Twenty years later, Spain was one of the most nuclearized countries in the world. A crucial role was played by US companies. The Spanish utilities decided to import US nuclear technology largely because of the financial facilities granted by US public and private institutions (Rubio-Varas and De la Torre, 2017).

In Slovenia there is the Krsko nuclear power plant, the result of a joint venture between Slovenia and Croatia (which at the time were part of the former Yugoslavia). The nuclear power plant was connected to the electricity grid in October 1981, but became operational only in January 1983.

In Italy, by contrast, much emphasis has been placed on natural gas, which accounts for about 60% of the production of electricity, and an energy basket dominated by the most expensive energy sources. Italy pursued a completely different energy policy, abandoning nuclear power in 1987. During the 1960s and 1970s nuclear schemes received considerable investments and met with great favour from the public at large, such that Italy gained a significant position in the world for the production of electricity of nuclear origin. The first power plants built were those of Latina (1963) and Garigliano (1964). With the entry into operation of the Trino Vercellese nuclear plant in Northwest Italy in 1965, the first phase of the industrial development of nuclear energy in Italy was completed. This expansive cycle was to be concluded with the activation of the Caorso plant in the Po Valley in 1980.

Italy's decision to build nuclear power plants, with the technologies available at the time, in the same period in which they were constructed in the USA and USSR, stemmed not only from the need to satisfy the growing demand for energy. In the middle of the Cold War with the Soviet Union, the

USA influenced the nuclear policies of European countries struggling with post-war reconstruction supported by the Marshall Plan. For the procurement of fuels for power plants, Italy (like Japan) would also be linked to an even greater dependence on the USA (Rubio-Varas and De la Torre, 2017). Yet in 1986, with the explosion of a reactor at the Chernobyl nuclear power plant in Ukraine, a critical consensus concerning nuclear energy arose. In Italy, implementation of part of the National Energy Plan, which provided for the opening of construction sites for new nuclear power plants, was blocked. In any event, with the 1987 referendum, Italy's electorate voted overwhelmingly to abandon the use of nuclear power as a form of energy supply. The referendum led to the closure of the nuclear power stations then in operation. After the end of its nuclear adventure, Italy focused heavily on natural gas, of which it is the largest consumer in the Mediterranean region.

In 2008, the Berlusconi government announced the resumption of nuclear power. The new nuclear programme envisaged the construction of four third-generation nuclear power plants by 2020, with a production of 6000 MW, in order to cover at least 10% of energy consumption in Italy. After the Fukushima disaster in 2011, Italy again abandoned its nuclear project through a new referendum, and decided to implement its policies in the gas sector, which imports from Russia and Algeria. Analysts believe that it was risky enough to focus so heavily on gas, since this means a strong dependence on its imports. This problem came to prominence on 1 January 2006, when the crisis began between Russia and Ukraine, a crisis that intensified again in January 2008 with the blocking of gas supplies by Russia. Relations between Russia and Ukraine have worsened following the 2014 referendum on the annexation of Crimea to Russia, and the tensions between the two countries are currently at the highest level.

In the Mediterranean region, renewable energy represents only 11% of total energy consumption, about three percentage points lower than the world level. Yet during the last decade, there has been significant progress in renewable energy within the Mediterranean region. Indeed, until 2009 renewables played a more marginal role, representing only 8% of the total. Changes in the energy balance of the Mediterranean region accelerated after the 2007 crisis. Renewable energy grew by 46% between 2007 and 2009. By contrast, oil consumption fell further, from 40% in 2009 to 37% in 2014, and 36% in 2016. This means that during the years when the effects of the crisis were felt more strongly in Europe and in the Mediterranean countries, the energy balance changed to the benefit of renewable energy, whose share has grown, while the overall contribution of fossil fuels decreased from 80% in 2007 to 76% in 2016.

It is also important to distinguish the energy balance within the single macroareas. In the Latin area, oil consumption increased from the 1960s, reaching its peak in the 1970s. Subsequently, in the 1980s, it suffered a sharp contraction and then settled on lower levels. After the economic crisis of 2007–2008, it again fell significantly. Yet the trend in other Mediterranean areas was different. In the Anatolian–Balkan area, in the Maghreb and, finally, in the Middle East, oil consumption has increased significantly since the 1970s. In particular, in the Maghreb, oil consumption has also risen in recent years. In the Anatolian–Balkan area, it has declined sharply since 2007. This decline has affected not only Greece, devastated by the serious economic crisis, but also Turkey, which has instead increased its consumption of coal. There was a sharp reduction in oil consumption in Libya due to events related to the end of the Gaddafi regime. After the peak of 14 540 ktoe in 2010, oil consumption plunged to 8981 ktoe in 2011. Then there was a recovery in 2012 and 2013, only to decline thereafter. In 2016, oil consumption stood at 10 717 ktoe. The chapters that follow investigate different energy balances by macroarea, with a special focus on $CO_2$ emissions.

## NOTES

1. IEA, Indicators for $CO_2$ Emissions, data extracted on 21 December 2018.
2. Namely: Algeria, Egypt, Israel, Jordan, Lebanon, Libya, Morocco, Palestine, Syria, Tunisia and Turkey.
3. IEA, Indicators for $CO_2$ Emissions, data extracted on 21 December 2018.
4. https://www.ecfr.eu/rome/post/laccordo_della_russia_con_la_turchia_sulla_siria_settentrionale, accessed on 10 January 2020.
5. IEA, *World Energy Balances*, 2018.
6. IEA, *Key World Energy Statistics,* 2018.
7. IEA, *World Energy Balances*, 2018.
8. European Parliament, 2014.
9. IEA, *World Energy Balances*, 2018.
10. IEA, *Key World Energy Statistics*, 2018. Data for coal production are 2017 provisional data.
11. IEA, *Key World Energy Statistics*, 2018. Data for nuclear electricity production refer to 2016.

# 2. Price dynamics, production and trade

## 1. OIL PRICE DYNAMICS

It is important to understand trends in oil prices in order to evaluate the macroeconomic scenario and possible implications for environmental policies, especially in the Mediterranean area, as well as the impact on the economy and political stability of oil-producing countries and the effect on importing countries. In early 2019, oil prices fluctuated between $60 and $65 a barrel, except between mid-February and late May when they ranged between $65 and $75 (Clô, 2019). Since 2016, the Organization of the Petroleum Exporting Countries (OPEC) has continued to pursue a policy of price stabilization, establishing a further reduction in production in January 2020. Moreover, since January 2020, the oil market has been hit by tensions between Iran and the USA, which arose following the American assassination of an Iranian general, Qasem Soleimani. Soon after, Brent crude, the international oil benchmark, rose by 3.5%, reaching $68.58 a barrel, and West Texas Intermediate, the US benchmark, rose by 3.2%, reaching $63.15 a barrel (Alabi, 2020).[1] Prices jumped soon after the assassination, with a spike of $70 per barrel, but after the initial reaction, they quickly eased, because there were no signs that Iran wanted to interrupt oil flows by closing the Strait of Hormuz. Differently from the past, oil prices now appear to react more slowly to tensions between the USA and Iran (Reed and Krauss, 2020). This confirms that not all price shocks are alike, as underlined by Fattouh and Economou (2019), who show that there is empirical evidence that supply and demand shocks do not have the same impact on oil prices in terms of duration and magnitude. With regard to supply shocks, such shocks can be exogenous, such as wars and other geopolitical tensions, or endogenous, due to decisions on production levels by oil-producing countries. As for demand shocks, these can be distinguished between flow demand shocks depending on the business cycle, and speculative demand shocks, the latter including precautionary demand as well. Empirical evidence has shown that flow demand shocks produce the largest and more persistent effect on oil prices. From 2000 to the eve of the 2008 international crisis, oil prices grew due to high demand, mainly fuelled by China and the emerging countries. This trend reversed with the 2008 crises, which caused a downturn in the business cycle at global level and a collapse of oil prices from June to December 2008. On the other hand, oil prices are not affected systematically by speculative demand shocks (Fattouh and Economou, 2019). On the supply side, endogenous supply shocks, such as the decision to cut oil production by producing countries, has a greater effect on oil prices, while exogenous supply shocks, linked to geopolitical tensions, have a major effect on prices but tend to be resolved in the short term, because the reduction in production of a country or group of countries is offset by higher production of other countries or by the use of precautionary reserves. During 2019, the Brent price increased, mainly due to cuts by OPEC, and especially by Saudi Arabia, and to a lesser extent due to geopolitical tensions (Fattouh and Economou, 2019), such as wars in Libya and Syria. It is far from easy to predict oil prices during 2020, not only because of continuing geopolitical tensions, but also because of the growth of trade barriers linked to tensions between the USA and China. Another important aspect is the relationship between gross domestic product (GDP) growth and spikes in oil prices. Hamilton (2011, 2013) has shown that 10 out of 11 post-war recessions of the US economy were preceded by a rapid increase in the oil price. He also found that the latest recession of 2007–2008 was preceded by a rapid increase in oil prices. Hamilton's analysis is empirical, and demonstrated a non-linear relationship between GDP growth and oil prices. On the other hand, Kilian and Vigfusson (2011) found little evidence of non-linearity in the relation between oil prices and US GDP growth, a different conclusion which led Hamilton (2013) to replicate that their analysis was based on a shorter period of time and different measures of oil prices and price adjustment. In addition, other studies also demonstrated a non-linear relation between oil prices and subsequent real GDP growth for a number of OECD countries (Cuñado and Pérez de Gracia, 2003; Jiménez-Rodríguez and Sánchez, 2005; Kim, 2012; and Engemann et al., 2011).

Oil prices rapidly recovered from the 2007–2008 global financial crisis. In 2010, Brent spot prices had soared to $93 per barrel by the end of the year. In April 2011, they rose to above $110. However, the political, social and economic situation rapidly changed with the spread of the Arab Spring in the Middle East and North Africa (MENA) countries, where citizens clamoured for reform and regime changes. In Tunisia and then in Egypt, demonstrations turned violent, leading to the end of long-standing regimes, such as that of Mubarak in Egypt and Zine El-Abidine Ben Ali in Tunisia. Conflicts erupted across the MENA region, from the Arabian Peninsula (Yemen) to the Levant (Iraq, Syria), and North Africa (Libya). As stressed by Jalilvand and Westphal (2018), energy plays a key role because various regimes legitimize their rule on the basis of energy, and in many of the MENA countries economic growth is based on the extraction and export of hydrocarbons. In addition, not only for producers, but also for importing countries, energy is an integral part of the social contract. At a distance of about three years from the beginning of riots in the Arab world, world energy markets have profoundly changed as a consequence of the strong decline in energy prices, from $114 per barrel in June 2014 to a minimum of $26 in January 2016, recovering to $55 at the end of that year. Jalilvand and Westphal (2018) argue that the decline in oil prices was caused by the ample supply at world level, due to several factors, such as technological development, which made it possible to produce "unconventional" oil and gas, whose supply grew especially in North America. At the same time, many oil producers, such as Saudi Arabia, expanded their production in a bid to win the competition on price. The growth of production was not followed by an equivalent increase in demand, due to slower economic growth in China and East Asia. The final effect was that oil production grew by a higher rate than demand, causing a dramatic fall in oil prices. Also Arezki and Blanchard (2014) suggested that, in the oil price slump between June and December 2014, factors on the supply side counted for more than factors on the demand side. In their explanation, a key role was played by OPEC's announcement to maintain current production levels despite the growth of oil production in non-OPEC countries, together with a faster than expected recovery of Libyan oil production and stable Iraq production despite the turmoil. By contrast, Baumeister and Kilian (2016) suggested that the 2014 decline in oil prices was mainly due to a demand shock resulting from unexpected worsening of the global economy.

As a reaction, OPEC and non-OPEC countries decided to cooperate, establishing new production policies with the aim of raising oil prices. They decided to reduce supplies in November 2016, reduction that was extended until March 2018, and extended again in 2019. Ample supply was limited not only to oil, but also affected gas markets, partly due to the shale gas revolution in the USA.

Another factor that contributed to changes in international energy was the growth of renewable energy production. Also in this case, one of the key factors was technological progress, which made it possible to achieve a significant reduction in costs of solar photovoltaics and onshore wind. In addition, global policies to reduce $CO_2$ emissions are pushing towards increasing production of renewable energy, thereby contributing to the increase in energy supply at world level.

Apart from the analysis of the factors that led to the oil price slump, our immediate concern is to highlight the turnaround in oil prices. From the energy crises of the 1970s to the riots of the Arab Spring, one of the main unsustainable factors of the energy system was the high price of oil. After reaching the historical maximum peak of $147 a barrel in July 2008, oil prices then plummeted below $40 as a result of the international economic and financial crisis. Between the end of 2009 and the first quarter of 2010, the price of crude oil rose to $80 a barrel, a price at which it stood for the whole of 2010. During 2011 it rose again, reaching an average of $110 a barrel due to the Libyan crisis and the growing demand from emerging countries. On the other hand, starting from 2014 the price of oil has plummeted, creating major problems for the producing countries, where the reduction in oil revenues translates into an increase both in the fiscal deficit and public debt.

The collapse in the price of hydrocarbons is having not only economic repercussions, but also effects in terms of balances between powers within the energy market. Saudi Arabia, Iran, Russia, Algeria and Libya, for example, need to sell their hydrocarbons to support their economies and are therefore dependent on their exports just as the Western countries depend on their imports. In this context, which has changed in just a few years, the governments of the various countries now need to examine the security of energy supplies, putting in place new policies that finally take note of the close interdependence between producer and consumer countries. The drop in oil prices could have greater social and economic consequences than the Arab Spring in 2011 (Bartoletto, 2016a).

## 2. ENERGY PRODUCTION IN MEDITERRANEAN COUNTRIES

Energy production in the Mediterranean region amounted to 546 446 ktoe in 2016,[2] corresponding to only 3.97% of total world energy production. From 1971 to 2007, in the Mediterranean region energy production doubled, but after reaching a peak of 666 384 ktoe in 2008, it started to shrink (Figure 2.1). During 2011, production levels fell by 12% compared to the previous year as a result of the Libyan crisis. In 2012 there was a significant recovery, but since then production has continued to decline. From 2008 to 2016, total production in the Mediterranean region decreased by 19.30%.[3]

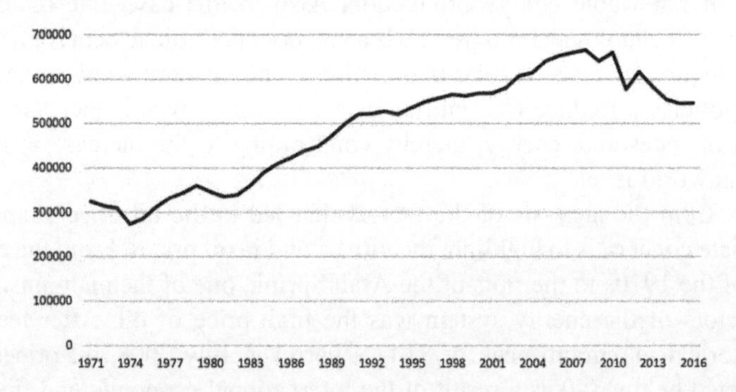


*Figure 2.1 Energy production in Mediterranean countries, 1971–2016 (ktoe)*

*Source:* IEA, World Energy Balances, data extracted on 19 April 2019.

The reduction has affected all areas of the Mediterranean region: from 2008 to 2016 energy production decreased by 54.2% in the Middle East, by 29.92% in North Africa, and by 1.16% in European countries (Table 2.1). In North Africa, all the countries have seen a reduction: in Libya this was very severe (−72.9%), but also Egypt experienced a significant contraction (−24.2 %), as did Tunisia (−20%). In the Middle East, the reduction was due to Syria, whose production plummeted (−82.5%), while in other countries, namely Israel and Jordan, production has increased.

In European countries, the picture is more variegated. Even though Malta has a very low energy production, production has rapidly increased in recent years (1872.4%). The same holds for Cyprus, which registered a positive growth rate (58.3%). In other countries, namely Portugal (34.2%), Spain (12.4%) and Italy (2.6%), growth has been less marked. The opposite occurred in other European countries, such as Greece (−32%), France (−3.8%) and Slovenia (−2.4%), production being lower in 2016 with respect to 2008.

*Table 2.1 Growth rates of energy production in the Mediterranean region, 2008–2016*

| Countries | Production 2016 | Growth rate % 2008–2016 | Countries | Production 2016 | Growth rate % 2008–2016 |
|---|---|---|---|---|---|
| **North Africa** | **257,825** | −29.9 | **Europe** | **220,319** | −1.2 |
| Algeria | 153,277 | −5.4 | Cyprus | 128 | 58.3 |
| Egypt | 67,615 | −24.2 | Croatia | 4,422 | −7.9 |
| Libya | 29,106 | −72.9 | France | 131,560 | −3.8 |
| Morocco | 1,783 | −5.5 | Greece | 6,707 | −32.0 |
| Tunisia | 6,044 | −20.0 | Italy | 33,770 | 2.6 |
| **Middle East** | **13,027** | −54.2 | Malta | 18 | 1872.4 |
| Israel | 8,274 | 111.6 | Portugal | 6,004 | 34.2 |
| Jordan | 355 | 28.7 | Slovenia | 3,585 | −2.4 |
| Lebanon | 176 | −5.3 | Spain | 34,125 | 12.4 |
| Syria | 4,222 | −82.5 | | | |

*Source:* Our calculations based on IEA, World Energy Balances, data extracted on 19 April 2019. Data on production are expressed in ktoe (kilo tons of oil equivalent).

Among the other countries of the Mediterranean region that are classified as European Union (EU) candidates, Turkey in 2016 produced 36 102 ktoe and registered a positive growth rate (25.85%) with respect to 2008. Albania also experienced a significant increase (70.88%), despite low starting levels of production.

As regards the distribution of total production of the Mediterranean region by area, about 47% is produced in North Africa, 40% in Europe, 10% in EU candidate countries and only 2.4% in the Middle East (Figure 2.2). Wars have modified the geography of production: prior to 2011, the contribution of North Africa and the Middle East to total production was higher: in 2010 North Africa produced 52.93%,[4] and the Middle East 4.85%.

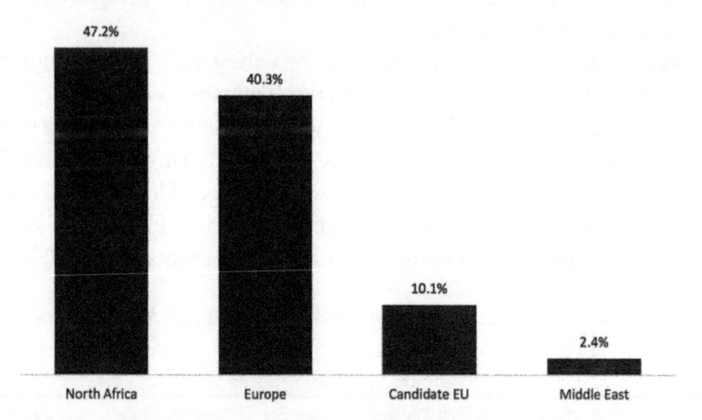


*Figure 2.2* *Energy production in the Mediterranean region by macro area in 2016*

*Source:* Our calculations on IEA, World Energy Balances, data extracted on 19 April 2019.

Even if production levels represent a small share compared to the world total and have declined over time, the Mediterranean plays a strategic role in the distribution of the main energy sources such as oil and natural gas due to its proximity to countries where important oil fields are located and to the presence of large oil and gas pipelines carrying the hydrocarbons of the former Soviet Union, or those of Iraq, Iran and other Gulf countries, to the markets of the West (Luciani, 1984; Keramane, 2001). If this is added to the availability of significant oil and gas reserves from some countries, above all Libya, Algeria and Egypt, but also Syria and Tunisia, the strategic role of the Mediterranean emerges with greater force, given that a considerable part of the world oil trade passes through the Mediterranean (Bartoletto, 2016b).

Diversity is the main feature of the Mediterranean area, which includes 25 countries that are very dissimilar in terms of economic size, levels of development and energy consumption. However, three categories of countries can be distinguished from an energy perspective: those heavily dependent on imports to meet their energy requirements (Portugal, Italy, Spain, Greece and France); those that produce and export significant quantities of oil and natural gas such as Algeria, Libya (until the 2011 crisis) and to a lesser extent Egypt; finally, countries that act as corridors for transit of energy products, such as Jordan, Israel, Lebanon, Egypt, Syria and Turkey.

Oil is a resource that has greatly influenced the stability of the Mediterranean region due to the growing role played in the economies of the industrialized countries since the end of the Second World War. The birth of the State of Israel in 1948 and its gradual strengthening in the territories of Palestine with the help of the USA, but against the will of the Arab countries, has caused serious military and diplomatic clashes and has led the exporting countries to use oil as a weapon of political and economic pressure against the West, in support of the Arab cause (Kapstein, 1990; Labbate, 2016). On the evening of 5 October 1973, the holiest day in the Jewish calendar known as Yom Kippur, the troops of Egypt and Syria invaded Israel. In the conference held in Tehran in December 1973, OPEC decided to increase oil prices by 130–140%. Following the rise to power of Ayatollah Khomeini in 1979 and the suspension of oil exports by Iran (then the second largest oil-exporting country in the world), the official average prices decreed and charged by OPEC between December 1978 and December 1979 increased from $12.8 to more than $26 a barrel, peaking a year later at around $45 a barrel, about 20 times the price ten years before.

The increasing consumption of oil in industrialized countries, the process of political emancipation of the Middle East and North Africa from Western hegemony, the reduction of unused production capacity worldwide, have all played a decisive role in triggering political tensions. Events such as the Yom Kippur War and later the rise of Ayatollah Khomeini to power in Iran in early 1979 led to an acute energy crisis, accompanied by a sharp rise in oil prices, which highlighted the key role that energy had assumed in the economy of modern societies and, at the same time, the great vulnerability of such societies due to their growing dependence on energy resources from politically unstable areas which were increasingly opposed to the values of Western cultures.

The five large countries of Southern Europe (Portugal, Spain, France, Italy and Greece) are heavily dependent on the Mediterranean countries that are part of the Libyan–Egyptian area and the Maghreb, from where substantial quantities of oil and gas are sourced. For this reason, during the Barcelona Conference of November 1995 the basic role of the energy sector in the Euro–Mediterranean partnership was recognized, with undertakings to strengthen cooperation in the field of energy policies, to create the conditions to encourage investment and allow energy companies to extend energy networks and promote interconnections (Balta, 2000). Fruit of a long negotiation process, the Euro–Mediterranean Conference laid the foundations for a global partnership to make the Mediterranean basin an area of dialogue, exchange and cooperation to guarantee peace, stability and progress (Bartoletto, 2016a). However, although the energy flows from the Mediterranean region may be significant, the latter, far from being self-sufficient in terms of energy, imports most of its needs from outside, especially from Russia and the Middle East.

Furthermore, the serious political, social and economic upheavals of recent years have had major repercussions on the energy market. We need only think of Libya, where the revolution that preceded and followed the end of the Gaddafi regime led to a sharp decrease in both oil and natural gas production. As a result, the supply of energy from the Mediterranean countries has decreased, and the energy deficit in the Mediterranean region has grown.

## 3. TRADE AND EXPORTING COUNTRIES

The Mediterranean plays a strategic role in the trade of energy sources for several reasons, such as the proximity to the major oil-producing countries in the Gulf such as Iran, Saudi Arabia, United Arab Emirates and Iraq. In addition, the Mediterranean region is crossed by important oil and gas pipelines that transport hydrocarbons from energy-producing countries, not only from the Gulf but also from the former Soviet Union, to the main supply markets, including Europe. The Mediterranean is also important due to the presence of countries producing oil and natural gas, such as Algeria, Libya and Egypt.

Libya is part of OPEC and, within the African continent, it is the country with the highest proven reserves of oil and has the fifth largest natural gas reserves.[5] The Libyan economy is very dependent on the export of hydrocarbons which, according to International Monetary Fund (IMF) data, represent more than 90% of government revenues. Libya produces significant quantities of high-quality crude oil containing a small amount of sulphur, which is exported mainly to European markets. Unfortunately, Libyan oil production, exports and related revenues declined sharply after the end of the Gaddafi regime, which was deposed in October 2011. In particular, during 2011 much of the production activity was suspended, resuming fairly rapidly in 2012, albeit not recovering to previous levels. In 2013, the oil sector was again hit by strikes and major unrest in the country, which led to extraction being halted in several oil fields and the blockade of oil terminals. The sharp reduction in Libyan exports also created difficulties for the refineries of the Mediterranean, which depend greatly on Libyan oil.

After a period of truce, in early April 2019 civil war broke out again around Tripoli with the advance of General Khalifa Haftar's troops. At the time of writing, oil and natural gas exports are once again at risk and Italy is the country that would be most affected by this new crisis. The recent epilogue of the war in Libya seems to be leading towards a division of its territories between Russia and Turkey.[6] In recent months, the involvement of Russia and Turkey in the Libyan crisis, also from a military point of view, has become considerable. Moreover, the path of Turkish intervention has been opened by the policies of the USA, which are ever less interested in the events of Libya, wishing to avoid major involvement, and Europe, which is struggling to find a unified position.

Turkey's interest in Libya is nothing new: for years Turkey has been, together with Qatar, the only country in the region to support the government of national unity (GNA) in Tripoli, in opposition to General Haftar. Also in this case, energy plays a key role, because there is the question of the exploitation of large gas reserves in Cypriot territorial waters. For years Erdoğan has been demanding, on behalf of the Turkish Republic of Northern Cyprus, recognized only by Turkey, joint exploitation of the gas resources, while the Republic of Cyprus, supported by Greece, claims its rights in what it considers its Exclusive Economic Zone (Talbot and Varvelli, 2019). Regardless of this, Turkey has already sent its drilling ships into Cypriot territorial waters, creating tension with Athens and Nicosia but also with the foreign oil companies operating in those waters. One of the aims of the alliance between Turkey and al-Sarraj is to define new maritime borders with Libya for an area that stretches from the south-western part of the Anatolian peninsula to the north-eastern coasts of Libya, especially the delimitation of the territorial waters around the island of Cyprus, to guarantee control over the huge gas resources that are found in those waters. Moreover, the interests at stake are even greater, as Turkey is seeking to affirm its role in the exploitation of gas reserves in the eastern Mediterranean. This contrasts with Egypt, Israel, Greece and Italy, which signed an agreement that at the beginning of 2019 established the Eastern Mediterranean Gas Forum for the joint development of hydrocarbons. Turkey therefore seeks the support of those actors, such as al-Sarraj, who are willing to support his plans in exchange for political and especially military support.

In Libya, after the peak in 1971, there was a sharp reduction in oil production and exports even before the two energy crises of 1973 and 1979. Following the second energy crisis, both production and exports settled on much lower levels than in the early 1970s. A recovery took place after 2002. Subsequently, first the international economic crisis of 2007–2008 and then the clashes that led to the fall of the Gaddafi regime caused a sharp reduction in oil production and exports.

Another aspect on which it is worth dwelling is the gap between production and exports. Until the late 1970s, the two curves overlapped substantially, given the very low levels of internal consumption. Subsequently, the gap grows, not only because production levels were lower, but also because domestic consumption increased. Libya exported most of its production to European countries. The main recipients were Italy, Germany and France. However, a significant change is emerging in the geography of supply, a change that is strongly affected by the emergence of new powers on international energy markets, such as China. While in 2008 only 4% of Libyan exports headed for China, in 2012 the share grew to 12%. Conversely, the Italian share decreased from 35% to 23% over the same period. In 2012, the other major destinations for Libyan exports were Germany (13%), France (10%), Spain (7%), Indonesia (6%), the UK (5%) and the USA (4%). From the data shown above, about 40% of Libyan exports headed for northern Mediterranean countries (Italy, France and Spain). If Germany and the other European countries are added, then about 70% of Libyan exports went to Europe, and only 25% to Asia. However, on comparing the 2012 data with those of 2008, the European share declined to the benefit of Asian countries, most of all China. During 2014, the geography of exports changed slightly, with the European share recovering. Indeed, about 84% of crude oil was exported to Europe, mainly to Italy, Germany and France. Libya also has important reserves of natural gas, and another priority objective was to increase its production and export. Through the Greenstream pipeline, Libyan natural gas arrives in Sicily and in the rest of the Italian peninsula and then continues to other European countries. Through this pipeline, 4640 billion cubic metres of gas reached Gela in Italy from Libya during 2017, corresponding to 6.7% of total imports.[7]

The pipeline is operated by ENI together with the National Oil Corporation (NOC), a large public company founded in 1970 that controls the oil and gas industry in Libya. Unfortunately, at the end of 2013 the situation in Libya flared up again and exports plummeted. Due to internal turmoil, despite having huge reserves of oil and natural gas, Libya was forced to buy diesel and fuel oil from neighbouring countries to keep power plants alive. By January 2013, Italy had already managed to import from Libya about 23% of its total imports, thus almost completely recovering to levels prior to the uprising that broke out in February 2011.

However, since then the situation has become precarious and while ENI, which was the first foreign energy company to operate in Libya, has stood firm, other companies instead, like the American multinational Exxon Mobil or the Anglo–Dutch Shell, have cut back their activity in the country (Bongiorni, 2013). During 2014 there was a recovery in oil production, but due to civil unrest, the situation again deteriorated. In 2015, production was still well below the levels of 2010. The Libyan economy was severely affected by the drastic reduction in oil production and exports. In 2011, the GDP collapsed and after recovering partly in 2012 it again fell about 14% in 2013 and 24% in 2014 following attacks and the partial destruction of the oil infrastructure.[8] To make up for the sharp fall in imports from Libya, Italy had to undertake the rather burdensome task of diversifying supplies, since oil is found on the spot markets, and therefore has a higher cost to which must be added the costs of oil tankers and insurance, which have significantly increased. Meanwhile,

Azerbaijan, which produces oil of a quality similar to Libyan crude, has established itself as Italy's main oil supplier. Libya therefore lost its leadership as the largest oil producer in the Mediterranean area in favour of Algeria, which is currently also the second largest oil producer and the largest gas producer on the African continent. Following the new unrest in Libya since 7 April 2019, not only oil exports but also gas exports are again at risk.

Meanwhile, even in Algeria the situation has become rather tense following the outbreak of popular protests against the president, Abdelaziz Bouteflika, who was forced to withdraw his candidacy in the elections that were postponed until the end of the year, keeping his power intact until that moment. If the crisis should escalate, Italy could suffer considerably, given that through the Transmed pipeline about 27% of total gas imports arrived in Mazara del Vallo in Sicily from Algeria during 2017.[9] Algeria is currently the most important energy-producing country within the Mediterranean region. Worldwide, in 2017, Algeria was ranked seventh among natural gas exporters, after the Russian Federation, Norway, Qatar, Australia, Canada and Turkmenistan.[10]

The Algerian economy is heavily dependent on the hydrocarbon sector which, according to IMF data, accounts for about 25% of GDP, more than 95% of export earnings and 60% of government revenues.[11] Sonatrach controls the production of natural gas in Algeria as well as its wholesale distribution, while Sonelgaz controls retail distribution. Hassi R'Mel is the main natural gas field, accounting for about 25% of total production. There are three gas pipelines connecting Algeria to Europe: the Trans-Mediterranean Pipeline, also named after Enrico Mattei, which starts from Hassi R'Mel and crosses Tunisia, Sicily and mainland Italy; the Maghreb–Europe Gas (MEG), also called Pedro Duran Farell, which starts from Hassi R'Mel, crosses Morocco and reaches Cordoba, where it connects to the Spanish and Portuguese gas networks. In July 2001 a consortium led by the Spanish company Cepsa and the Algerian company Sonatrach signed an agreement to build a new pipeline, Medgaz, which starts from Beni Saf in Algeria and arrives directly in Spain, without going through Morocco. This new pipeline has been operational since 2011.[12]

Algeria is also an important oil exporter and a member of OPEC. Since the 1970s there has been an increase in production volumes, with a slowdown following the international crisis of 2007–2008. Sonatrach handles not only natural gas but also oil production, transportation and marketing. The gap between oil production and consumption remains very wide, even if in recent years the growth in domestic consumption has led to a reduction in the surplus to be exported. Saharan Blend, produced in Algeria, is one of the best quality oils in the world. Precisely because of its very low sulphur content, it is in great demand in European countries, where regulation on the contents of petrol and diesel is becoming increasingly stringent.

Up to 2012, about 45% of Algerian oil exports were shipped to Europe and about 36% to North America. The USA represented one of the main export markets for about a decade until 2013. Subsequently, US oil imports fell substantially, and as a result the role of Europe as the main outlet market has grown.

Until the 1980s, Egypt was a major oil producer and exporter, a role that it subsequently lost. Since 2010, oil consumption has even exceeded production, and exports have decreased significantly. From this perspective, Egypt plays an important role in the energy market thanks to the Suez Canal, which connects the Red Sea and the Gulf of Suez to the Mediterranean Sea. To better understand the strategic role of Egypt, it is worth pointing out that about half the world's oil production is traded by sea, which is why the Suez Canal plays a fundamental role for energy security and has been the protagonist of major disputes in the past, as we will see in depth in the next chapter. The canal is a major transit point for ships travelling northwards, from the Persian Gulf to Europe and North America, on the one hand; on the other, for southbound vessels, that is, from the North African producers to Asia. A strategic role is also played by the Suez–Mediterranean pipeline, known more commonly as the Sumed pipeline, which connects the Red Sea to the Mediterranean through northern Egypt. The Sumed pipeline is the only alternative for transporting crude oil to Mediterranean countries should ships be unable to use the Suez Canal.

In 2014, a large offshore gas field was discovered in Egypt by ENI, which is very active in the country. The discovery of this new field is of great importance. In the first place, it strengthens the role of Egypt as a regional power. On the economic side, it will attract new investments, generating jobs and making Egypt more self-sufficient in terms of energy and even able to export gas. The Zohr field is extraordinary, not only for its size and the quality of gas but above all for its location, as it is situated a short distance from the target markets and with infrastructure and gas treatment and liquefaction plants already largely in place. This discovery confirms ENI's international leadership in the exploration of hydrocarbons. However, the discovery is not expected to greatly reduce Italy's dependence on Russia, given that the larger amount of gas produced will be mainly used to satisfy Egyptian requirements. However, the political tensions within the country are slowing down operations.

Syria was a major producer of oil and gas in the Eastern Mediterranean in the past, but over the years, production and exports have progressively diminished. Oil and gas production have fallen dramatically since civil war broke out in spring 2011. The energy sector has been affected not only by the conflict, but also by the sanctions imposed by the USA and the EU.[13] Energy infrastructures such as oil pipelines, gas pipelines and power grids have been severely damaged.

Before the current conflict, the energy sector provided about a quarter of government revenues. The main oil fields of Syria are on the border with Iraq. From 2008 onwards, hence already before the revolution, Syria became a net importer, but the ongoing war and sanctions have compromised imports. Natural gas is imported through the Arab Gas Pipeline (AGP), which allows Egyptian gas to be sent to Syria via Jordan. The pipeline has been the subject of frequent attacks and is currently out of order.[14] However, Syria continues to play a strategic role from an energy standpoint because it is close to the oil pipelines and sea routes that carry most of the crude oil. To the southwest is the Suez Canal, which connects the Red Sea and the Gulf of Suez with the Mediterranean. Another regional oil route potentially exposed to the Syrian crisis is the Sumed pipeline which, as mentioned above, transports oil from the Persian Gulf region to the Mediterranean. To the north of Syria is Turkey, which is crossed by important oil and gas pipelines. To the east are the oil fields in the northern part of Iraq, a major oil producer (Philbin, 2013).

Turkey is a corridor that connects the Black Sea with the Mediterranean, the main outlet point for Asian oil pipelines. The country therefore plays an increasingly important role in the transit of oil, precisely because it is strategically located between the rich producers of the former Soviet Union and the Middle East on the one hand, and the European importing countries on the other.[15] Thus it may be said that the strategic importance of the Mediterranean is not limited to the presence of major energy-producing countries, but concerns more generally the geopolitics of energy. The Turkish straits (Bosporus and Dardanelles waterways) represent a bottleneck, through which about 2.4 million barrels per day of crude oil and petroleum liquids were transported in 2016.[16] The Bosporus is a 17-mile waterway that connects the Black Sea with the Sea of Marmara. The Dardanelles is a 40-mile waterway that links the Sea of Marmara with the Aegean and Mediterranean Seas.

In addition, Turkey is crossed by strategic pipelines, namely the Baku–Tbilisi–Ceyhan (BTC) oil pipeline, which transports oil from Azerbaijan and Kazakhstan, and the Kirkuk–Ceyhan pipeline carrying much of Iraq's oil to the Mediterranean. The Iraqi portion of the pipeline was the target of militant attacks and stopped operating in 2014.[17]

Turkey also plays a strategic role for natural gas because it connects continental Europe with the large gas reserves of the Caspian basin and the Middle East. The Blue Stream Pipeline, whose project was designed by Gazprom, ENI and Botes, transports Russian gas to Turkey via the Black Sea; the Iran–Turkey Pipeline, inaugurated in 2002, transports natural gas from Tabriz in Iran. Other gas pipelines are being expanded, such as the South Caucasus Pipeline, also known as Baku–Tbilisi Erzurum (BTE), completed in 2007, which allows Turkey to import gas from Azerbaijan. Greece and then Italy will also import gas from Azerbaijan through Turkey, thanks to the Turkey–Greece–Italy

Interconnector. Although work on the Turkey–Greece interconnector started in 2007, the extension of the line through Greece and Italy is proceeding very slowly.[18]

Meanwhile, Turkish President Erdoğan and Russian President Putin met on 8 January 2020 in Istanbul for the inauguration of the Turkish section of TurkStream, the pipeline that will bring Russian gas to Turkey and Europe from the southern shore of the Black Sea in the coming years, thus bypassing Ukraine. Once completed, this gas pipeline, whose construction began in 2017, will represent a real turning point in the energy supply market and hence in the energy geopolitics of the Mediterranean region. Indeed, the Bulgarian Prime Minister Boyko Borissov and Serbian President Aleksandar Vucic also took part in the inauguration ceremony.

The increase in consumption and growing instability in the Middle East are driving European countries towards greater diversification of supply areas and towards consolidation of import quotas. Russia is once again emerging as an energy superpower, using its oil and natural gas resources as a political weapon. A significant share of European gas imports come from Russia and, according to the most recent estimates, this percentage is expected to rise. Russia is conducting a policy designed to broaden its sphere of influence in the Mediterranean, and in particular North Africa.

In Libya and Syria, Russia and Turkey are backing rival parties. On the Libyan front, Turkey is keen to support Fayez al-Serraj's GNA in Tripoli, which controls the west of the country, while Russia is backing its rival, Haftar's eastern-based Libyan National Army (LNA).[19] Continued European and American absence helped increasing influence of Turkey and Russia who seem to prefer Libya's split into a Russia-influenced east and a Turkish-dominated west.

The Algerian state energy firm Sonatrach and the Russian Gazprom, Europe's leading gas suppliers, erstwhile competitors, have signed an important agreement for natural gas (with effect from 2006) and, more recently, oil. Sonatrach in 2014 made a new oil and gas find with Russia's Gazprom following successful drilling in the North Berkin Basin in the Algerian Desert, and in previous years other joint discoveries were made.

However, there are projects that, on the contrary, weaken the role of Moscow on world energy markets, such as the BTC pipeline, strongly supported by the USA with the aim of weakening the links between Russia and the former Soviet republics of Azerbaijan and Georgia. The BTC pipeline is the first oil pipeline to transport oil directly from the Caspian Sea to the Mediterranean without passing through Russian territory or the crowded Bosporus strait. It was inaugurated on 13 July 2006 in the Turkish city of Ceyhan, in the presence of the Presidents of the three states affected by its passage, namely Turkey, Azerbaijan and Georgia, as well as senior US officials. The 1765 km-long oil pipeline is expected to supply over 50 million tons of crude oil annually, accounting for 6–7% of the world oil flow. Not only Russia, but also the countries of the Middle East, Iran in particular, are weakened by the opening of the above pipeline, the fruit of US policy to counteract Moscow and at the same time reduce dependence on supplies from the Middle East, whose instability generates major concerns in relation to energy security. By contrast, the new oil pipeline reinforces Turkey's role as an energy bridge between East and West. In the next chapter, we will focus on historical aspects of the role of energy in international politics and alliances after World War II.

## 4. MIDDLE EASTERN COUNTRIES

Procuring energy resources has always been a difficult issue for Israel, engaged in continuous conflict with its neighbours. Over the years, Israel has sought to become a petro-state like some of its neighbours, but attempts to find oil have been unsuccessful. The reduction of energy dependency has always been a priority in the Israeli political agenda, and for this reason, not only has domestic oil exploration been carried out, but also several attempts have been made to introduce renewable sources in the energy mix. This picture completely changed after the discovery of two large offshore natural gas fields in Israel's Exclusive Economic Zone (EEZ), respectively the Tamar gas field in 2009, and Leviathan in 2010.[20] These discoveries were seen as a divine gift, being a major geopolitical asset for the energy independence of Israel. In addition, the new gas fields gave Israel the opportunity to export gas to its neighbours (Rettig, 2018).

To better understand the impact of gas discoveries, it should be borne in mind that, until the 1980s, Israel was completely dependent on oil, most of which was imported from Iran. The 1973 OPEC embargo and the Iranian revolution of 1979 led Israel to replace oil with coal to satisfy its energy demand. Natural gas was introduced to produce electricity only in 2004, after the discovery of Noa and Meri, two small offshore gas fields close to the city of Ashkelon. In addition, Israel imported gas from Egypt, through a pipeline that started to operate in 2008. However, the two small offshore gas fields were rapidly depleted, and Egyptian imports had been affected both by the continual sabotage of the pipeline, and the rapid reduction in Egypt's export capacity due to its higher domestic demand. By 2012, imports from Egypt had ceased, and Israel's electricity sector suffered a gas shortage, which was offset by more expensive oil and additional coal, with an increase in the energy bill.

Given the long-term energy instability of Israel, one can understand why the discovery of the large gas fields of Tamar and then Leviathan, which is almost twice as large, had major consequences for the country. Later the two fields of Tamar and Leviathan were supplemented by the discovery of two smaller fields, Tanin in 2012, and Karish in 2013. Thanks to these discoveries, the role of natural gas in Israel's energy balance has rapidly increased, replacing coal, while oil continues to maintain a large share. The large gas reserves represented the opportunity not only to satisfy domestic electric demand, but also to export to neighbouring countries, such as Jordan and Egypt, even though the discoveries led to new maritime border disputes with Lebanon.

The discovery in 2015 of the Zohr gas field in Egypt, and from 2014, the fall of oil prices, and globally low gas prices have impacted Israeli gas policy exports, because the cost of producing gas and transporting it by pipeline makes this option commercially less convenient (Ellinas, 2018).

Within the Eastern Mediterranean, not only Israel and Egypt, but also Cyprus, have experienced the discovery of a gas field (Aphrodite in 2011), with a probable reserve of 127 billion cubic metres (Ellinas, 2018). However, while the economics and geopolitics of gas and energy are transforming the Eastern Mediterranean, political instability and frequent upheavals make it more complex to attract investment from companies and banks.

Lebanon is an important energy bridge between Western Europe and Middle Eastern oil-producing countries, and its connection with oil has a long history. A starting-point was marked by the 1928 Red Line Agreement, a deal signed between Britain, France and the USA, with the aim to find oil in the Middle East, within territories of the former Ottoman Empire. This agreement was a fundamental step for oil development in the Middle East, Lebanon included, because it granted the Iraq Petroleum Company (IPC) the licence to explore for oil in Lebanon, close to Jabal Terbol, six miles from Tripoli. From then on, several attempts were made to find oil within Lebanese territory, as in 1947 by the Lebanon Petroleum Company, but they were unsuccessful, and Lebanon did not become a petro-state (Atallah and Fattouh, 2019). However, Lebanon maintained a strong relationship with oil because it was an important transit for Iraqi and Saudi oil on its way to Western markets. As early as 1934, oil was transported from Kirkuk in Iraq to Tripoli through the IPC pipeline. After World War II and for the next 50 years, Lebanon reinforced its oil transit role because in 1947 it agreed to host the Trans-Arabian pipeline (Tapline), which transported Saudi Arabia's oil to Western Europe through the port of Saida. Lebanon thus not only became an energy bridge between Middle East oil-producing countries and Western Europe, but also became part of US policy influence (Atallah and Fattouh, 2019). It should not be forgotten that the Trans-Arabian pipeline was constructed thanks to funds from the Marshall Plan.

However, Lebanon has not abandoned its desire to become a petro-state. After about 70 years from the 1947 unsuccessful attempt to discover oil, Lebanon has renewed its hope to become a petroleum-producing country after the discovery of new oil fields in the Eastern Mediterranean made by Cyprus and Israel. For this reason, the Lebanese government commissioned surveys in Lebanon's coastal waters, and in 2010 the first Offshore Petroleum Resources Law (OPRL) came into effect. In addition, the Council of Ministers issued a decree that established a petroleum authority (LPA), but the process came to a halt in August 2013 due to political paralysis. The negative effects of the delay in developing the sector have been amplified by the sharp reduction in oil prices, since few companies have expressed interest in applying for an exploration licence.

Natural gas has played a very limited role in Lebanon's energy mix due to a lack of access to gas supplies. Higher international oil prices from around 2005 encouraged Lebanon to reconsider the role of gas, since it could save a large amount on its annual fuel bill if it switched its power generation from oil to gas (Fattouh and El-Katiri, 2019). Moreover, falling oil prices since 2014 have made the energy picture even more complex.

The main constraint to expanding the share of gas in the energy balance has been access to gas supplies. Natural gas entered the energy mix of Lebanon for the first time in 2009, when the AGP, which already supplied Jordan, started supplying Egyptian gas to the Beddawi power plant. It was a very brief experience, because from the beginning the Egyptian gas supplies were often interrupted due to delays in payments and a number of explosions that hit the AGP. The last supply of Egyptian gas to Lebanon was made in November 2010, and also Jordan was affected by several delivery cuts. The rise in gas consumption in Egypt and the unstable political situation strongly reduced supplies to regional partners. In 2003, Lebanon signed a 25-year contract with Syria to import natural gas, through the Gasyle pipeline, which extends from the Syrian border to the Beddawi power plant, and was completed in 2005. However, Syria was unable to supply natural gas to Lebanon, because its own domestic demand exceeded production. At present, the ongoing war in Syria, as well as Lebanon's changing relations with its neighbours, make it impossible to make predictions on the country's energy trade.

Like Lebanon, Jordan is also dependent on energy imports, despite its proximity to very energy-rich countries, such as Iraq and Saudi Arabia. Jordan's energy balance is dominated by fossil fuels, which in 2016 represented about 97% of total energy consumption. The main energy source is oil, with a share of about 56%, followed by natural gas, which meets about 38% of total primary consumption, while renewable energy sources represent only 3% of total consumption. In recent years, gas imports have experienced several problems and Jordan has been affected by shortages in power generation. Until 2010, natural gas represented about 80% in electricity production, but when in 2011–2012 imports from Egypt plummeted, the share of natural gas in electricity production fell to 20% (Abu-Dayyeh, 2018). Thus it is evident why ensuring a stable energy supply is one of the main priorities in the Jordan political agenda, but this aim is strictly connected with political stability, which currently appears assured despite its proximity to countries experiencing wars, such as Iraq and Syria, or political crisis, such as Egypt and Lebanon. Although Jordan has a parliament, effective political power is in the hands of the Royal Court. During the Arab Spring, Jordan also experienced protests, whose trigger was the government's decision to reduce subsidies, despite increasing unemployment. To end the protests, the country had to review its economic policies: incentives were reintroduced, including those related to energy consumption. The latter, together with an energy dependency of more than 90%, contributed to the worsening of fiscal imbalances. From 2000 to 2013, when oil prices were rising, the cost of energy imports grew from 7.5% of GDP to 16.3%, contributing to the rise of public debt (Abu-Dayyeh, 2018). The decline in oil prices since 2014 has had only a minor impact on public balances, because it has been outstripped by the rise in energy consumption. Hence the cost of energy imports remains very high.

## NOTES

1. Alabi (2020), available on https://www.ft.com/content/1155034d-6d5b-4388-bfa3-0c51760a01be, last accessed on 14 January 2020.
2. IEA, *World Energy Balances*, data extracted on 19 April 2019.
3. Our estimation based on IEA, *World Energy Balances*, data extracted on 19 April 2019.
4. Our calculations based on IEA, *World Energy Balances*, data extracted on 19 April 2019.
5. US Energy Information Administration, Country analysis brief: Libya, 19 November 2015.
6. https://www.arabnews.com/node/1605086/middle-east, last accessed 17 February 2020.
7. For data on imports by point of entry, see https://dgsaie.mise.gov.it/gas_naturale_importazioni.php.
8. US Energy Information Administration, Country analysis brief: Libya, 19 November 2015.
9. For data on imports by point of entry, see Ministero dello Sviluppo Economico (2018), accessed at https://dgsaie.mise.gov.it/gas_naturale_importazioni.php.
10. IEA, *Key World Energy Statistics*, 2016.
11. US Energy Information Administration, Country analysis brief: Algeria, 11 March 2016.
12. US Energy Information Administration, Country analysis brief: Algeria, 11 March 2016.
13. US Energy Information Administration, Syria, full report, last updated 24 June 2015.
14. Arab Gas Pipeline (AGP), Jordan, Syria, Lebanon, https://www.hydrocarbons-technology.com/projects/arab-gas-pipeline-agp/, last accessed 25 February 2020.
15. US Energy Information Administration, Country analysis brief: Turkey, 1 February 2013; 7 July 2015.

16. US Energy Information Administration, World oil transit chokepoints, 25 July 2017.
17. US Energy Information Administration, Country analysis brief: Turkey. Last updated: 2 February 2017.
18. US Energy Information Administration, Country analysis brief: Turkey. Last updated: 2 February 2017.
19. https://www.arabnews.com/node/1605086/middle-east, last accessed 17 February 2020.
20. Ellinas (2018) argues that the Tamar gas field has probable reserves of 280 billion cubic metres (bcm), and Leviathan has probable reserves of 620 bcm. Rettig (2018) argues that Leviathan holds about 500 bcm, while Tamar has about 282 bcm.

# 3. Past and present of energy security in Mediterranean countries

## 1. ENERGY SECURITY SINCE THE WORLD WARS

Energy security in Mediterranean countries has been a major concern since the early twentieth century when military needs dictated the switch from coal to oil: on the eve of World War I, the British government opted for oil as the source of power for the Royal Navy upon pressure exerted by Winston Churchill, First Lord of the Admiralty. This meant that the British fleet would no longer be reliant on British coal but on oil imported from far-away Persia (Kapstein, 1990). The link between energy and politics further strengthened during World War II. Economic and security issues were inextricably linked in the Western alliance. Middle Eastern oil was considered an important factor for the success of the European Recovery Programme (Anderson, 1981). This meant that European countries reduced their reliance on domestic coal and became dependent on oil from the Middle East.

Energy also became an important factor in the "Cold War". The USA was the first to promote an alliance in the energy sector. Indeed, energy policy was one of the main aims of the Western alliance post World War II, a fact which is often overlooked. Prior to the energy crises of the 1970s, the West was hit by four fuel crises in 1944, 1951, 1956 and 1967. Indeed, energy security was a major concern of the Western alliance throughout the second post-war period. Gilpin (1975) and Kapstein (1990) argue that the monopoly of nuclear weapons in 1945 and the control of oil and monetary reserves were the three main sources of power that enabled the USA to establish the post-war order.

Maintenance of adequate energy supplies at reasonable cost was a common interest to Western allies, since energy had become fundamental for military, industrial and civilian requirements. At the end of World War II, Europe suffered severe coal shortages, which was tantamount to a serious energy crisis since at that time coal was the world's main energy source. Before World War II, coal accounted for about 90% of Europe's energy demand, being indispensable for transportation, home heating, industry and the generation of electricity. During the war the largest European mines had been destroyed or exhausted. Thus Europe would require huge coal imports. The coal shortage in Europe and relative import requirements were the focus of attention by the USA: they could produce this amount of exportable surplus, but their loading capacity and the fleet of cargo ships were insufficient. The solution was found in the creation of the European Coal Organisation (ECO), which aimed to coordinate European demand for coal and mining equipment, give technical assistance, identify priorities and so forth. The ECO was founded in London on 18 May 1945 by Belgium, Denmark, France, Greece, Luxembourg, the Netherlands, Norway, the UK, the USA and Turkey. The Soviet Union did not join the ECO, because it had not reached an agreement with the USA. Following the Soviet decision, East European countries did likewise (Kapstein, 1990). The institution of the ECO, whose primary task was to allocate coal to member states, shows that energy and international politics were closely linked and that energy played a role of primary importance in international relations. Later it became clear that it was impossible for Europe to recover pre-war levels of coal production despite the creation of the European Coal and Steel Community (ECSC) in 1951. Europe would no longer be self-sufficient in energy terms. The energy shortfall resulting from diminishing coal supplies steered firmly towards Middle Eastern oil.

The Marshall Plan had an important impact on Europe's energy policy. During the years 1948–1951, the USA supplied those European countries which were beneficiaries of the plan with extensive aid for fuel purchases, although the amount of aid for oil purchases far exceeded that for coal. Hence the impact of decisions taken within the Marshall Plan significantly changed the geography of energy supplies: Europe began to be reliant on oil from the Middle East, reducing the level of energy autonomy which in previous years had been afforded by domestic coal production. In addition, the USA discouraged Western Europe from importing Polish coal. The resumption of coal production in the Ruhr was an important aim of the Marshall Plan. Most of the capital invested in the Ruhr was supplied by the Economic Cooperation Administration (ECA), which was founded in Washington to supervise the Marshall Plan. In December 1948 the International Authority for the Ruhr (IAR) was established, comprising, France, the USA and the Benelux countries. The main aim of the IAR was to control the distribution of coal, coke and steel from the Ruhr. Its institution represented an important shift of energy policies leading ultimately towards the Cold War. While action undertaken by the ECO distributed coal to all member states, including Poland and Czechoslovakia, with the IAR the energy policy changed completely: the issue of coal allocation concerned only the Western alliance (Kapstein, 1990). While before the war, Poland was a major coal exporter, it subsequently lost this role as the Western alliance wanted to minimize dependence on Soviet bloc producers. These new "Cold War" strategies, together with a real difficulty to restore pre-war levels of production, accelerated the transition to oil.

With the shift from domestic coal supplies to oil, the problem of energy security became much more pressing. It was immediately clear that the Middle East was an unstable area, and that oil flows could be stopped by internal and external factors alike. Since World War II, also the Soviet Union had sought to extend its influence over the Middle East, especially Turkey and Iran. The political situation in Palestine complicated the picture, since Palestine served as the terminus of several Middle East oil pipelines and was an important refining centre in the region. The Marshall Plan gave oil a major impetus. Indeed, the ECA was entrusted by Congress to assist the overseas operation of US oil companies, which needed steel, equipment to drill and build pipelines, refineries and port facilities in the Middle East.

An important project that was started within the Marshall Plan was the construction of the Trans-Arabian Pipeline, also called Tapline, which was to carry crude oil from Saudi Arabia to the Mediterranean. This project was conceived because under the US strategy more oil was needed for the recovery of Western Europe. The construction of new routes to the Mediterranean had become more important since the Palestine conflict led to the closure in early 1948 of the oil pipelines to Haifa.

The initial Tapline project involved the construction of a pipeline running over 1000 miles along the northern border of Saudi Arabia, across Transjordan, and then either through Palestine or through Syria and Lebanon. Yet the construction of the oil pipeline was of greater importance to the USA within the context of the Cold War. To gain an insight into the issue, a step backwards needs to be taken. In May 1933, Saudi Arabia granted the Standard Oil Company of California a concession to find oil, which led to the discovery of oil in 1938. In the midst of World War II, as it was clear to government planners that the USA would become a net importer after the war, they were keen to ensure continuing access to Saudi oil and rapid development of Saudi reserves for the benefit of the USA and the Western alliance. In 1948 Standard Oil discovered the world's largest oilfield in Saudi Arabia, namely Ghawar. To conclude, within the Marshall Plan, policies for security of energy supply played a key role. US government strategy supported the expansion of Middle Eastern oil with the aim of preserving domestic supply, also by curtailing exports, for energy security reasons. Yet this strategy had several weaknesses, there being many age-old internal conflicts in the Middle East which constituted a threat for oil flows. An important source of tension was the conflict between Arabs and Jews. The establishment of the State of Israel aggravated American relations with Arab countries and, to a much greater extent, complicated the management of a strategy of dependence upon Arab oil.

In 1947, the Special Committee on Palestine created by the UN proposed the partition of Palestine, which was accepted by the General Assembly of the UN and the Jewish Agency, but was rejected by the Arabs. On 14 May

1948, the Jewish National Council proclaimed the State of Israel (Yergin, 1991). The reaction of the Arab League was very hostile, and resulted in the first Arab–Israeli war. But while Jews and Arabs were fighting in Palestine, oil production continued with Saudi Arabia. The construction of Tapline also continued, which became operative from November 1950. Thanks to this new pipeline, oil arrived at Sidon in Lebanon, the terminus on the Mediterranean, where it was loaded by oil tankers to be transported directly to Europe. Tapline replaced a much longer sea journey from the Persian Gulf through the Suez Canal. Once again the Mediterranean region confirmed its strategic role in world energy markets.

## 2. THE FIRST POST-WAR OIL CRISIS (1951–1952)

Before World War I, oil production in Persia was insignificant. In 1912 it amounted to only 1600 barrels per day. During World War I, Persian oil production grew more than tenfold. In 1918 it was about 18 000 barrels per day. In 1916, Persian oil satisfied a fifth of the British Navy's oil demand (Yergin, 1991). At the end of World War II, the UK had large economic and political interests in Iran and, on the basis of an agreement with Washington, it assumed responsibility for defence of the Middle East in the event of a global conflict with the Soviet Union. Thus, even though the USA diverged from the UK in several matters of policy with respect to Iran, in the end it supported the British government in resolving the dispute between the Iranian government and the Anglo–Iranian Oil Company (AIOC) about agreements made before and during World War II concerning the sharing of profits between oil firms and governments of oil-producing countries (Kapstein, 1990). The British government was a major shareholder in the AIOC and many British workers of the AIOC lived in Iran. The AIOC was an important oil supplier of Britain and significant quantities of oil were shipped to Western Europe.

However, tensions were running high because nationalists considered the concession of AIOC nationalization a legitimate right of the Iranian people. In the meantime the American Oil Company Aramco and Saudi Arabia had signed a fifty-fifty profit-sharing agreement that reinforced the idea in Iran, Egypt and other countries, of being exploited by Britain, ultimately weakening its position in the Middle East region. Events precipitated and on 28 April 1951, Muhammed Mossadegh, leader of the nationalist movement, became the new Prime Minister of Iran, and nationalized the AIOC, taking away from the latter any control over the oil industry. Riots broke out at the Abadan refinery, two British employees of the AIOC were killed, and many British workers left Iran. After the Abadan riots, the AIOC reduced its activity and by the end of July 1951 the refinery had shut down.

It was a blow, most of all for the UK, but also for other Western European economies as the loss of Iranian oil forced a rapid change in the geography of energy supplies. Iranian crude oil was replaced with oil from Kuwait and Saudi Arabia, but major problems arose with the supply of refined oil products which had hitherto been produced in the Abadan refinery, the largest in the world. Another major problem was the transportation of oil products. Furthermore, the Iranian oil crisis led to a rise in oil prices. With the closure of the Abadan refinery, Europe had lost its main supplier of jet fuel and fuel oil. As a result, the new energy security policies tended towards increasing European refinery capacity. In particular, Italy expanded its refining activity, becoming Europe's refinery by the 1970s (Bartoletto, 2005).

In the meantime the Iranian economy had collapsed, partly due to the international boycott of Iranian oil and the resulting currency crisis. Crude oil production in Iran fell from 19 million barrels in June 1951 to zero by September (Hamilton, 1983).

Nationalization of the AIOC also led to tensions between the UK and USA, since the latter wanted to maintain its leadership in the Western alliance, curb the rise of communism and confirm its hegemony in the Middle East. On the other hand, there was mutual interest between Iran and the UK in maintaining a strong partnership: Britain had important economic and political interests throughout the Middle East and in formulating a political response it had to consider that the moment was particularly delicate since Egypt was seeking British military withdrawal from the Suez Canal Zone.

In August 1953, with British and American backing, Mossadegh was arrested and the Shah returned to Tehran, establishing a pro-Western political regime. From an energy security perspective, this allowed the USA and Britain to secure Iran's oil supplies once again. But it was clear that the growing dependence of Europe on Middle Eastern oil created enormous problems for the security of energy supplies. Between the end of 1953 and the beginning of 1954 a new agreement was reached between Iran, the USA and Britain. The Iranian Consortium was established to produce, refine and sell Iranian oil, with a fifty-fifty oil agreement between foreign oil companies and the Iranian government.

The National Iranian Oil Company (NIOC) became the owner of all assets which had previously been held by the AIOC. After the signing of the consortium agreement, the Abadan refinery resumed its activity under NIOC supervision. Thanks to the agreement between Hoover[1] and the Iranian government, Western markets had oil supplies guaranteed for 20 years.

## 3. THE SECOND POST-WAR OIL CRISIS: THE SUEZ CRISIS (1956–1957)

The Suez crisis of 1956–1957 represented another major shock for international oil markets, since about 70% of oil exported from the Middle East to Europe was shipped through the Suez Canal, which connects the Mediterranean to the Red Sea. The nationalization of the Suez Canal in November 1956 led to the interruption or at least a bottleneck of oil flows, hence an oil shortage in Europe. Not only in Iran, but also in Egypt, nationalist sentiments were emerging. The Egyptian government asked for British troops to withdraw from its Canal Zone, where troops had been allocated following a pre-war agreement. In particular, after the Italian invasion in Africa during the Fascist period, Britain and Egypt in 1936 signed an Alliance that allowed Britain to deploy its troops in Egypt. The treaty was to last twenty years, until 1956, and could be reviewed only by mutual agreement. At the end of World War II, the Egyptian government asked for the treaty to be renegotiated, but the British government, which was responsible for Middle East defence, had no wish to reduce the deployment of troops significantly in the Suez Canal, under the justification that the Suez Canal could be one of the Soviet Union's main objectives in the event of hostilities, a concern shared by the USA. In October 1951 the crisis erupted: Egypt repealed the treaty of alliance and the UK refused to accept this unilateral decision. As a consequence, fighting broke out between the two countries. Egypt was plunged into a climate of terror that contributed to weakening the existing government. In July 1952, a coup d'état ended the monarchy of King Farouk and ushered in a new military government. Behind the coup was General Nasser. The new military government resumed negotiations with Britain and on 27 July 1954 an agreement was signed between the two parties.

Under the leadership of Prime Minister Nasser, an agreement was finally signed between the Egyptian and British governments, under which British troops would have a month to leave the Suez Canal Zone. However, in the event of war, the agreement provided for the return of British troops. Through political, military and economic support, Britain had played an important role in the Middle East, in countries such as Iraq, Jordan and Egypt. Yet the departure of British troops represented a significant weakening of the UK's position in the Middle East.

After the signing of the agreement, the USA provided Egypt with a loan for its economic development. Nevertheless, after US support for Israel's creation in 1948 and the resizing of the English position in the Middle East, Nasser's Egypt began to open up significantly to the Soviet Union. First, Nasser announced that on the basis of an exchange agreement, Egypt would receive arms from Czechoslovakia and in return it would send Egyptian cotton to the Eastern bloc. In addition, Nasser announced in 1955 that he wanted to build a large dam at Aswan, and that the Soviet Union had promised financial support. In the end it was an expedient to obtain other finance. Indeed, the first phase of construction was financed especially by the World Bank.

However, Nasser's relations with the West further deteriorated when it was announced that the Soviet Union had offered to build an atomic power station in Egypt. On 26 July 1956, President Nasser announced that the Suez Canal Company would be nationalized, and that its revenues would be used to finance the Aswan High Dam. It should be pointed out that the Suez Canal was opened in November 1869, and although the canal lay in the sovereign territory of Egypt, it was owned and operated by the Suez Canal Company, which was a private company owned by British and French shareholders. The largest single shareholder was the British government, while the headquarters of the Suez Canal Company were in Paris. Tolls were paid directly to the Suez Canal Company, which gave the Egyptian government only a share of the profits. The concession was to last until 1968, and on expiry could be renegotiated. But Nasser decided unilaterally on new terms. It came as a shock because of the large quantities of freight moving through the canal. Initially, the Suez Canal was of strategic importance for Britain as it afforded a swift passage to India. After India's independence in 1947, the canal maintained its strategic importance for world trade. Indeed, most of the oil from the Middle East was transported to Europe via the Suez Canal. At that time there were only three routes for channelling Middle Eastern oil towards Europe, namely the Suez Canal and two pipelines.

The nationalization of the Suez Canal was a source of great concern for Britain, unlike the USA which relied on the canal for oil supply only in limited quantities. Thus the positions of Britain and the USA on Nasser's policies differed. France was also opposed to Nasser's policies, not only due to the Suez Canal issue, but also because Nasser was lending military support to Algerians rebelling against the French control of their country. In order to limit the growth of the influence of Nasser in the Middle East region, France provided Israel with advanced military equipment. Yet, as stressed by Kapstein (1990), Nasser had nationalized and not confiscated the Suez Canal Company, promising that compensation would be paid to shareholders and that the flow through the canal would not stop. Indeed, Egypt needed the canal revenues to finance the construction of the Aswan Dam and other plans to promote economic development. While the UK and France sought military intervention, the USA preferred a diplomatic solution, also to avoid the risk of increasing the influence of the Soviet Union in the area.

The issue of the Suez Canal was complicated by the tensions between Israel and its Arab neighbours: terrorists attacked Israel, gaining access from Jordan and Gaza. Nasser was considered a common enemy by Israel, France and Britain, which planned a military strategy against Egypt.

On 29 October 1956 Israel launched its attack with French military support. On 31 October France and the UK launched an air strike against Egypt, hitting Egyptian airfields. As a reaction, on 2 and 3 November, Egypt sank blockships in the Suez Canal, halting the flow of traffic. In addition, the Iraq Petroleum Company's pipeline that passed through Syria was damaged. As a consequence, oil flows to Europe were severely disrupted and Europe experienced a real energy crisis. To make matters worse, Saudi Arabia blocked oil exports to the UK and France.

Clearly, Europe needed to find new oil-supplying countries and new routes for oil transportation. With the loss of the Suez Canal, the oil tankers carrying oil from the Persian Gulf to Western Europe had to travel a much longer route, circling the Cape of Good Hope. Indeed, the Suez crisis resulted in a major change in the geography of oil supply, with the growth of oil imports from Texas and Venezuela.

By the end of May 1957, with the re-opening of the Suez Canal and the Iraq Petroleum Company (IPC) pipeline, the oil crisis was over. Yet it was clear to Western European countries that they needed to diversify their energy supplies, both in terms of energy sources and supplier countries. Growing concerns with the security of energy supplies pushed Europe towards nuclear power as an alternative source of energy. As early as 1955, countries that were part of the ECSC decided to create EURATOM, a supranational atomic energy organization. The British government, which had not participated in the EURATOM project, announced during the Suez crisis an important project of construction of nuclear power plants. After the Suez crisis, Western European countries began to buy oil from the Soviet Union. Sweden was the first country to sign a bilateral oil agreement with Russia, as a reaction to the Suez crisis. It was followed by Austria, Italy, West Germany and France. Indeed, in 1962–1963 the Kennedy administration increased pressure on Western European countries to stop buying Soviet oil, because since 1956 Soviet oil exports to Western Europe had increased by more than 700% (Jentleson, 1984). The USA feared that through the supply of oil, the Soviet Union could extend its political influence to such countries. Italy attracted special attention: it not only imported about 22% of its oil from the Soviet Union, but had a strong Communist Party in Parliament (Jentleson, 1984). Yet the East–West oil trade was not only an issue involving Cold War ideology, but it also concerned large American oil multinationals, which saw a sizeable share of their profits taken away by the Soviet Union. Germany was also buying oil from the Soviet Union, to ensure cheaper, diversified supplies. Indeed, Soviet exports to Western Europe rapidly increased after price cutting. Until 1957, the Russian oil price had been higher than its Arabian counterpart, which was why imports from the Soviet Union grew, but in a more constrained manner. However, in 1958 and 1959, when the Soviet Union started to undercut Arabian oil, exports rapidly increased (Campbell, 1968).

The oil market had profoundly changed, and an important contribution to this change was made by Enrico Mattei, one of the main protagonists of the Italian economic miracle, and the founder and first President of Italy's National Hydrocarbon Authority (ENI). In 1945 he was appointed extraordinary commissioner of the Italian General Oil company (AGIP), a public company founded in 1926 during the fascist regime to ensure Italy's autonomy in energy provision and which dealt with prospecting, refining and the distribution of hydrocarbons. During the war, Mattei became a partisan leader, and by April 1945 was considered a leading figure of the new ruling class, so much so that the role of AGIP Commissioner seemed unsuitable (Colitti, 1979). AGIP had not achieved substantive results; it was considered by many to be too close to the old fascist regime and hence was to be wound up as soon as possible. However, Mattei had been informed by AGIP technicians that there were a series of indications that revealed the presence of natural gas and oil reserves in Caviaga, in the heart of the Po Valley. Mattei immediately grasped its importance, urging geologists and drillers to intensify their search. Thus instead of liquidating AGIP, Mattei immediately took action to find the financial means necessary to start drilling the Caviaga wells. He also informed the prime minister that in his opinion AGIP had to be saved. To carry through his project, he waged a tough battle against the sceptical politicians, the old leaders of AGIP, the private industrialists and the foreign oil companies, with a view to entrusting a state-run company with the monopoly of the search for, and distribution of, hydrocarbons in the Po Valley.

For Mattei, however, the Po Valley monopoly was just a starting point for a much larger project to supply Italy with energy, which also saw the construction of a dense network of pipelines. Mattei's ultimate aim was to ensure that Italy had its own energy supply and was no longer dependent on British and American companies. His first move was to establish ENI in 1953, a large state-run authority. He was very critical of the large oil companies, which he referred to as the *Sette Sorelle* (Seven Sisters). He was aware that in the meantime, after the Suez crisis, British influence in the Middle East had waned and that this was the right time to deal directly with the Middle East to split oil revenues. Mattei began his negotiations with Iran and the Shah. For his part, the Shah needed to earn higher profits from Iranian oil than he obtained from the Consortium with the Seven Sisters. The new deal signed in 1957 between the NIOC and ENI represented a watershed for world oil markets.

ENI's penetration into Iran deserves attention because it radically changed the geopolitics of energy both in the Mediterranean and elsewhere. But let us take a step back. In the mid-1950s Mattei initiated a completely new relationship with the governments of oil-producing countries. He realized that the countries concerned wanted to appropriate the wealth of their subsoil, until then managed according to the rules and prices defined by the great international oil companies. When ENI was established, the seven major oil companies (Standard Oil California, Standard Oil New Jersey/ Esso, Gulf, Texaco, Mobil, BP and Shell) controlled more than 90% of world oil reserves outside the USA, Mexico and Communist countries, 90% of production and 75% of refining capacity, thereby accounting for 90% of oil sales on international markets. This allowed them to exercise informal control of the world's oil economy while keeping prices at very profitable levels. Until 1959 Italy had paid higher prices than other Western European nations even though it was closer to the Middle East and should have benefited from lower transport costs. The success of Mattei's political initiatives with Arab and African countries emerging from the colonial experience derives precisely from the courageous and unexpected act of breaking the oil rules made with the allocation of royalties. Mattei succeeded, in a few years, in establishing personal friendships with many heads of state who held power in Arab and African countries at the end of colonization after World War II: the King of Morocco Mohammed V, the Egyptian Gamal Abdel Nasser, the Tunisian Habib Bourguiba, the Algerian Ahmed Ben Bella and the Shah of Persia Reza Pahlavi who Mattei met several times in Rome and in Metanopoli. Mattei offered the above countries the possibility of becoming operators in a joint venture with ENI–AGIP, with the consequence that profits would no longer be distributed on a historic fifty-fifty basis but would achieve the 75:25% split in favour of the energy producers, the so-called Mattei formula, which led to a revolution in the world oil market. Naturally, the Western oil giants reacted adversely and sought to prevent at all costs the penetration of ENI into Africa and other Arab countries.

Much of the behaviour of the major companies *vis-à-vis* ENI was directly inspired by the governments of the countries from which their shareholders came: the USA, UK, the Netherlands and France. There were concerns, especially among Americans, that Italy, through Mattei, could become a pawn in the strategy of Soviet penetration into the "Third World". After the Iranian revolution and the nationalization of the oil industry, the AIOC had been replaced by the NIOC. Nationalization was followed by a period of crisis, as the AIOC closed its installations on Iranian soil, leaving the country devoid of competent structures and technicians, hence unable to exploit its deposits. The British then implemented an embargo against Iranian oil, which effectively prevented its sale to major oil companies worldwide. A compromise was reached in 1954 with the creation of a consortium with large international companies, which obtained in the same year a very large concession on Iranian territory, based on a 50% profit-sharing agreement. The consortium was in fact made up of the most important oil companies in the world. Despite indicating its willingness to join the consortium, ENI was overlooked and thus started to collaborate independently with the NIOC. Mattei believed that the negotiations for the formation of the consortium were of an essentially political nature, and that Italian companies were excluded in principle. Mattei then looked for an alternative channel to Iran to the consortium. Iranian oil policy was heavily influenced by the consortium, which had a huge concession in the most fruitful areas of the country. Not having the means to change this state of affairs, the Iranian government aimed to enhance the areas left free, focusing mainly on collaboration with Italian, German and Japanese groups, with which to set up joint ventures. The countries in question, having very few oil resources, would conduct the research operations with greater commitment than that demonstrated by the consortium, and in the Iranian government project they would also have a rebalancing effect on the political life of the country, excessively influenced by the economic dominance of the Anglo-Americans. A petroleum agreement with Iran, under the terms established in August, was broadly considered beneficial for both countries: it would first ensure Italy had oil resources without a large outlay of currency; Iran would then constitute a huge outlet for the export of machinery and materials, and there were finally promising opportunities to make advantageous agreements also for the extraction and treatment of natural gas. Conducted by Enrico Mattei with the Shah and his ministers, the negotiations ended with the final ENI–NIOC agreement on 14 March 1957, stipulated by Mattei himself in Tehran.

The agreement provided for the creation of a company with Iranian and Italian joint ownership, SIRIP, which would carry out oil prospecting in three areas covering 23 000 $km^2$. The new oil agreement required several months to be approved. It led to the Petroleum Act, which entered into force on 8 September 1957. The US government opposed the agreement and attempted to have it cancelled (Alfieri, 2019). Even though Mattei died on 27 October 1962, the world oil market had changed, and the rapid growth of oil exports from Libya contributed to that change.

## 4. OPEC, LIBYA AND THE THIRD POST-WAR OIL CRISIS

In the meantime, other important changes had occurred on international energy markets. In 1960, during a conference organized in Baghdad, the Organization of Petroleum Exporting Countries (OPEC) was created with the aim of acquiring full control over oil production and prices. The founding of OPEC represented the response of oil producers to falling oil prices on international markets. The OPEC countries were concerned about the reduction of revenues from their oil exports, with the lowering of prices affected all oil-producing countries. Had this been limited to one country, the government of that country could have been compensated by an increase in its share of world production. Being instead generalized, it constituted a powerful incentive for the governments of the producing countries to respond with one voice. Among the countries that promoted the establishment of OPEC, there were not only the Arab countries, but also others such as Venezuela, which was one of the main promoters (Luciani, 1976). The founding states were: Iran, Iraq, Kuwait, Saudi Arabia and Venezuela. The five founding states were later joined by other members: Qatar (1961), Libya (1962), the United Arab Emirates (1967), Algeria (1969), Nigeria (1971), Ecuador (1973), Angola (2007) and Gabon (1975).[2]

In response to growing oil dependence on OPEC, in 1961 the Organisation for Economic Co-operation and Development (OECD) was founded, replacing the previous Organisation for European Economic Co-operation (OEEC), created to administer Marshall Plan aid, which in the meantime had exhausted its role, the USA having completed its aid plan for the reconstruction of Europe.

Moreover, the new OECD presented important novelties, because it included Canada and the USA, which joined as member states. A few years later, New Zealand and Japan also joined. The OECD was based only on voluntary cooperation between member countries, unlike the OEEC, which imposed rules that member states had to respect. One of the main issues on which the OECD started to work concerned oil dependence on OPEC. In Western Europe about 95% of oil demand was satisfied through imports, while Japan imported 100% of its energy requirements. About 80% of the imports came from the Middle East and North Africa.

Since the early 1960s Libya has become one of the leading oil-exporting countries. The exploitation of oil resources has radically changed the economy of the country but also produced major socio-political changes. Yet in 1951 Libya had political sovereignty but little else beside. It was one of the world's poorest countries. The battles of World War II had destroyed what infrastructure had been built and disrupted the economic life of even the Bedouin communities. Italian colonization in 1935 failed to push the country towards modernization and the construction of an education system. Although Libya was an Air Force base, one of the main American bomber bases in the Eastern Hemisphere, it received little consideration at international level. The country was so poor, plagued by droughts and locusts, that its economic prospects were very negative (First, 1974).

Nobody knew that Libya possessed large oil reserves. However, at the beginning of the 1950s, geologists suspected that there might be oil in the subsoil, so in 1955 the first Libyan Petroleum Law was enacted. Interestingly, from the very beginning Libya sought to follow a different route with respect to the countries of the Persian Gulf. Indeed the new law provided only for several small concessions and not for one or a few large concessions as in Iraq, Saudi Arabia or Kuwait. The aim was to prevent one oil company being able to control the entire country, as explained by the then oil minister (First, 1974). The law provided for many concessions to independent oil companies. Nevertheless, work was far from easy: on the oil fields there were still many mines left over from World War II. Indeed, at the beginning of oil exploration, many workers died from exploding mines. The first discovery was made in 1958 by Esso Standard at Atshan in Fezzan, while a major discovery took place a year later in Zelten, where large fields of low-sulphur crude oil were found. By 1961 crude was being produced and exported. Oil revenues began to course through the economy on a staggering scale from 1963 onwards. Thanks to oil exports, Libya became not only financially self-reliant but wealthy enough to influence other countries. With the 1969 revolution, Gaddafi and a group of young officers ousted the monarchy. Gaddafi was gripped by a vision of the need to develop his country, transform its society, and regenerate and unite the Arab world (First, 1974).

The increase in oil production and exports was impressive. Libyan oil had several advantages, because it was low-sulphur and therefore more suitable for the production of oil products such as petrol, whose demand was rapidly growing together with the number of cars. This was an important aspect: in that period there was growing concern for the environment, which pushed towards the use of Libyan oil which was a low-sulphur crude, less polluting than heavier crude from the Persian Gulf. Another major advantage regarded its location, since Libya was very close to Italy's refineries and in geopolitical terms it was not in the Middle East: Libyan oil did not pass through the Suez Canal.

The vast surge of Libyan oil had a strong impact on world oil prices, because it contributed to a declining phase. Between 1960 and 1970 the oil market price decreased significantly. At least half of the production was in the hands of independent oil companies. Before oil was discovered, the Libyan economy was based almost exclusively on agriculture and pastoralism. After the discovery of oil, agriculture lost its role as the main sector of the national economy. Furthermore, the public administration also played an important role. Oil revenues became the main source of entry for the Libyan government. In 1966, the Libyan government received about nine million Libyan pounds from taxpayers and about 138 million Libyan pounds from oil companies. The following year, taxpayers contributed about ten million Libyan pounds while oil companies accounted for 170 million (Luciani, 1976). This means that the population could afford to be indifferent as to how public money was used, because an increase in public spending did not lead to an increase in taxation thanks to oil revenues.

Yet exploitation of oil fields is an activity that requires the intervention of the government of the country in which the oil fields are located. The state should divide exploration rights, regulate extraction and transport, and establish with oil companies the conditions of concessions. Thus a link is created between oil production and political life. As oil revenues became the main state income, the regulation of the oil sector became a fundamental role for the Libyan State. Thus oil changed not only the economy but also Libyan politics. Libya's reserves were huge, its oil quality was particularly sought after because of its low sulphur content, it was closer to Europe and unaffected by events such as the closure of the Suez Canal or Tapline attacks. The oil fields are located in the middle of the country, on the border between Tripolitania, Cyrenaica and Fezzan. If oil had only been discovered in one of the three regions, the history of Libya could well have been different (Luciani, 1976). Moreover, until oil was discovered, it was not important for Libya to be so active in support of the Arab cause, as it was a poor country. But after the discovery of oil, the Arab countries, above all Egypt, pressed for Libyan support for the Arab cause. On 1 September 1969, while the King was abroad, a group of young officers took power in a bloodless coup. Thus King Idris was deposed and Senussi power quickly dissipated. Gaddafi's rise to power was not the only shock to international oil markets: there were also previous events, such as the Six-Day War between Israel on one side and Egypt, Syria and Jordan on the other, as part of the long-running Arab–Israeli conflict, which resulted in rapid and total victory for Israel. On 5 June 1967, Israel launched its first attacks against Jordan and Syria. On 6 June, the day after fighting began, the Arab oil ministers decided on an oil embargo against countries friendly to Israel. Thus Saudi Arabia, Kuwait, Iraq, Libya and Algeria prohibited oil exports to the USA, Britain and, to a lesser extent, West Germany (Yergin, 1991). For the first time, the Arab countries had coalesced to use oil as a weapon. The Iraq Petroleum Company pipeline and the Tapline (Trans-Arabian Pipeline) were blocked. In addition, once again the Suez Canal was closed and it was again necessary for oil tankers to re-route. With the Six-Day War, the Mediterranean, Europe and the World experienced the start of the third post-war oil crisis.

## 5. A WATERSHED IN MEDITERRANEAN ENERGY SECURITY: THE FOURTH OIL CRISIS (1973–1974)

The Arab oil embargo of 1973–1974 represented a watershed in terms of security of energy supply in the Mediterranean area. To find out why, we need to focus on important changes that occurred in the world economy during the twenty years before.

The 1950s and 1960s were years of great economic growth, during which industrialized countries became richer and other countries that lagged behind could be transformed into industrialized nations, such as Italy and Japan. Such economic growth was mainly fuelled by oil, whose world demand rose from 19 million barrels per day in 1960 to more than 44 million barrels per day in 1972 (Yergin, 1991). Oil became the main energy source in the industrial sector, and was also used for new domestic requirements. Indeed, during this period, life-styles in many countries had radically changed. The number of cars had risen sharply in many countries, which resulted in a considerable increase in fuel consumption. In the USA alone, the number of motor vehicles increased from 45 million in 1949 to 119 million in 1972 (Yergin, 1991). During the 1950s and 1960s, the price of oil fell and became very cheap, which contributed to the massive growth of oil consumption. In the meantime, surplus capacity in the USA had significantly diminished, since rising demand had outpaced new oil field discoveries. From 1957 to 1963, surplus capacity in the USA was about 4 million barrels per day. By 1970, this had declined to only about a million barrels per day (Yergin, 1991). This was an important issue in terms of energy security, because the USA needed to import growing quantities of oil from other countries. Indeed, oil imports rose from 19% in 1967 to 36% in 1973. The reduction in the US security margin also had important repercussions for energy security policies of Western European countries. Meanwhile, as production in the Middle East and North Africa had grown, industrialized Western countries were able to satisfy their rising demand.

In the meantime, the debate on the environmental impact of growing energy demand was climbing up the political agenda. The main conclusion was that coal consumption had to be reduced in favour of less polluting oil. Thus environmental policies prior to the 1973–74 oil crisis did not consider an expansion of renewable energy sources, but simply a switch from coal to oil. Another important consequence was the expansion of nuclear as an energy source. Already during the Six-Day War, the idea of oil at the centre of the energy system had begun to waver: the Suez Canal had closed during the war, and remained closed until 1975. In the meantime, Iraq's exports had declined due to disputes between the government and the Iraq Petroleum Company. Besides, in May 1970, an accident at the Trans-Arabian Pipeline disrupted exports to the Mediterranean area. Already these events were causing major changes in the structure of the world's oil supply, favouring new actors such as Gaddafi's Libya. An important advantage of Libya was its proximity to Europe, especially to Italy and other European Mediterranean countries.

Oil consumption had grown exponentially in OECD countries, putting pressure on the issue of energy security. When the new Arab–Israeli war broke out in 1973, the world oil economy changed radically. The Six-Day War had concluded with the rapid victory of Israel, which gained a dominant position in the Middle East. The discontent in the defeated Arab countries was high, especially because Israel had conquered new territories in Sinai and the West Bank of the Jordan. In 1969 conflict broke out once again between Israel and Egypt along the Suez Canal, and relations between Israel and Palestine became increasingly tense, with continuous dramatic terrorist attacks.

In 1973, Egypt's President Sadat decided on war, bolstered by the support of other Arab countries which had shared the problem of liberating the occupied territories. In addition, Sadat was convinced that European countries would not support Israel, being highly dependent on oil from the Middle East and North Africa. Like Nasser before him, Sadat believed in oil as a lever against the Western world, including the USA. At this point, the equilibrium within the Mediterranean area played an increasingly important role for the security of energy supply at world level. Since security margins in the USA had diminished, they could not provide support to their Western allies in the event of an oil embargo.

On 6 October 1973, during the Jewish holiday of Yom Kippur, Egypt and Syria invaded Israel. It was a great success for Arab states. During the following days, large oil flows through Middle East pipelines were stopped. Egypt and Syria benefited from a massive arms shipment from the Soviet Union. In reaction, US President Nixon authorized arms shipments to Israel. In Kuwait, 16 October saw a meeting of the Organization of Arab Petroleum Exporting Countries (OAPEC). During the meeting, the Arab oil ministers decided to raise prices unilaterally, from $3.011 to $5.119 per barrel. The day after, they decided to reduce oil production until Israel had withdrawn from the Arab territories occupied during June 1967 and restored the rights of Palestinians. Concerned for their energy security, France and Britain refused to support Israel, supplying arms instead to Arab countries. France for example supplied weapons to Libya and Saudi Arabia which then transferred the arms to Egypt and Syria (Kapstein, 1990). In short, this new Israeli–Arab conflict divided the Western alliance. On the one hand, the USA supported Israel while, on the other, the European countries maintained a more favourable attitude to the Arab countries for fear of an oil embargo. On 18 October, Abu Dhabi and Algeria announced a complete embargo of the USA. In the following days other oil-producing countries also declared an embargo of the USA, which was extended to the Netherlands, also guilty of supporting Israel. Other European countries such as France, Spain and the UK did not suffer any kind of penalization and received full oil supplies because they were considered friendly countries. Neutral states, such as Germany, suffered a small reduction in their supplies, but well below 5% previously declared by the Arab oil ministers within OAPEC. The increase in the price of oil was decided by the producing countries partly to defend their economy from inflationary processes, also triggered by the devaluation of the US dollar at the beginning of 1973

(Luciani, 1976). In 1971, after Nixon ended dollar convertibility into gold, the Bretton Woods System was brought to an end, and not only the USA, but also European countries were affected by high inflation rates. Despite rising prices, European countries continued to purchase oil not so much to satisfy consumption as above all to accumulate stocks. There was great uncertainty in oil world markets and nobody knew how long the oil embargo would last or how severe the oil shortage would be. Oil ministers met in Tehran in late December 1973 to decide the official price of oil. After a heated discussion, the new price was fixed at $11.65 a barrel. Thus the oil price had risen from $1.80 in 1970 to $2.18 in 1971 to $2.90 in mid-1973 to $5.12 in October 1973, and finally to $11.65 in December 1973. In the end, Iran's hard line had prevailed (Yergin, 1991).

The oil embargo lasted officially until 17 March 1974. This crisis showed that the USA no longer represented the energy supplier of last resort. During and soon after World War II, the USA played a leading role in the Western alliance also in energy security terms, because they were considered the energy supplier of last resort. After the Six-Day War, the USA lost this role, and Western Europe followed a different strategy in the Middle East, because it "cared more about oil than about Israel" (Kapstein, 1990, p. 170). However, the USA remained the main economic and military power within the alliance. Moreover, the oil price rose and it was clear that the crisis would not be short-lived. The era of cheap oil had ended.

As a response to the Arab oil embargo, in 1974, under pressure from the USA, a special agency was created for energy within the OECD, the International Energy Agency (IEA). But not all OECD members accepted to participate in the IEA. Indeed, France, Finland and Iceland refused: France due to its pro-Arab policies (although the IEA began its operations in Paris, at the headquarters of the OECD), and Finland and Iceland because they were largely dependent on oil imports from the Soviet Union. In reality, the Yom Kippur war accelerated deep-rooted changes that had occurred in the oil market even before, with the rise to power of Gaddafi in Libya in 1969. The end of long-term agreements increased oil price volatility because short-term prices respond rapidly to political and economic events.

Moreover, after the oil embargo, oil demand had diminished, which was a concern for OPEC countries, since industrialized countries in the meantime had tried to reduce dependence on OPEC oil by switching to oil from the North Sea and Alaska's North Slope.

Four years after the Yom Kippur war, concern over another Arab–Israeli war seemed to be dispelled as the Arab leader Sadat visited Israel on 20 November 1977. Secret negotiations were held at Camp David (Maryland, USA), with an agreement signed in the White House by the Egyptian President Anwar Sadat and the Israeli Prime Minister Menachem Begin on 17 September 1978, under the auspices of the then US president Jimmy Carter. The agreements led to the 1979 Israeli–Egyptian peace treaty and the withdrawal of Israeli troops from Sinai. Following the agreements, Egypt was suspended from the Arab League from 1979 to 1989. Thereafter the oil-producing countries followed different strategies. Saudi Arabia nurtured its relations with the USA, while Algeria and Libya tended towards isolation (Kapstein, 1990).

## 6. THE FIFTH POST-WAR OIL CRISIS (1979–1981)

The worst crisis for the oil markets was about to come: the Iranian revolution of 1978–1979 caused panic and a rapid increase in oil prices. Starting from World War II, there had been massive growth in world oil consumption. From 1949 to 1972, world energy consumption had more than tripled, while oil consumption increased more than five-fold. During the same period, in the USA oil consumption tripled, from 5.8 million to 16.4 million barrels per day, in Western Europe it rose by a factor of about 15, from 970 000 barrels per day to 14.1 million barrels per day, and Japan's growth was quite staggering, consumption rising 137-fold from 32 000 barrels per day to 4.4 million barrels per day (Yergin, 1991). Coal, which had powered the Industrial Revolution during the eighteenth and nineteenth centuries, had lost its primacy in the energy system. Therefore, during the 1970s the world was dependent on the Middle East and North Africa for oil, and there was the perception that the twenty-year surplus in oil markets had ended. The two key oil-producing countries of the Middle East in the 1960s and 1970s were Iran and Saudi Arabia. The world battle for oil production in general and the growth of oil production in the Persian Gulf in particular intensified the rivalry between these two countries for the allocation of production growth. But despite the rapid increase in oil revenues, Iran remained a very backward country. The urban population grew but agricultural output was declining and food imports were rising. Infrastructures were poor and the national electricity grid was unable to satisfy energy demand. Tehran and other cities often experienced black-outs, even for several hours in a day, which was a disaster for industrial production and a great inconvenience for everyday life. Galloping inflation worsened the already difficult economic and social situation. At a political level, the Shah's regime had lost consensus, favouring the opponent Ayatollah Ruhollah Khomeini, which considered the Pahlavi regime corrupt and illegitimate. The Shah was unable to restrain the growing rebellion. Much of the country, including the oil industry, went on strike during September and October 1978. The oil industry was in the most complete chaos. The strikes spread to the "Fields", the main production area in Iran. Their impact was felt immediately, since Iran was the world's second largest exporter of oil, after Saudi Arabia. Exports had greatly diminished. On 13 October 1978 the Abadan refinery was shut down.

On the political level, the causes of the Iranian crisis are very complex. A major role was played by Islamic fundamentalism that was firmly set in opposition to the Western world. By December 1978 Iranian petroleum exports had ceased, and by the end of the month the Shah would leave Iran. The Pahlavi dynasty was finished and the long-time opponent of the Shah, Khomeini, returned to Tehran on 1 February 1979. The old regime was gone in Iran, and a new one was in power. Up to 1978 Iran had been the world's second-largest oil exporter, after Saudi Arabia. The new oil shock had several phases, the first from the end of December 1978, when Iranian oil exports ceased, to autumn 1979. The loss of Iranian production was partially offset by increases in Saudi Arabia production and from other OPEC countries. In the end there was only a small loss of oil supply: about 4–5%. Although the reduction in oil supply was marginal, it resulted in a 150% price increase. The main cause was panic, fuelled by the rapid growth in oil consumption from 1976 and the disruption of contractual arrangements within the oil industry after the Iranian revolution.

Shortly after, another threat for energy security emerged: the Soviet Union's invasion of Afghanistan on 27 December 1979. The USA considered the Soviet move towards the Persian Gulf of great concern. However, US policy regarding the Middle East in general and Afghanistan in particular was not altogether shared by its allies, Western Europe and Japan. The problem lay in the security of energy supply. Indeed, after the oil crisis of 1973, Western European countries and Japan had stepped up their relations with the Soviet Union for energy supplies and in 1978 had already begun negotiations for the construction of a pipeline that would bring natural gas from Siberia to Western Europe.

In September 1980, Saddam Hussein, Iraq's dictator who rose to power in 1979, declared war on Iran. Iraq invaded and the Iranian Abadan refinery came under siege. The Iranian counterattack caused damage to Iraqi oil installations and pipelines. The Iran–Iraq war, which ended in 1988, led to significant oil removal from world markets. However, the longer-term response in demand from consuming countries to the price hikes of the 1970s was quite considerable, and world petroleum consumption declined significantly in the early 1980s. Indeed, the period 1981–1986 witnessed a great price collapse. The price of oil collapsed from $27 a barrel in 1985 to $12 a barrel at its lowest point in 1986 (Hamilton, 2011b). Oil production from Iran and Iraq was very slow to recover its pre-war levels.

## 7. PERSIAN GULF CRISIS AND SIXTH POST-WAR OIL CRISIS

The Persian Gulf crisis represented the first oil crisis to occur after the end of the Cold War. When Saddam Hussein's Iraq invaded Kuwait, international markets had another jolt, since the new war became a geopolitical oil crisis, and the oil question was one of the reasons for the anti-Iraqi reaction. A new order in international energy markets was needed, especially after the fall in 1989 of the Berlin Wall, which had symbolized the Cold War since the end of World War II. On 31 December 1991 the Soviet Union collapsed and was replaced by the Russian Federation, together with a group of newly created republics. With the collapse of the Soviet bloc, the regional balance of power shifted, and the state of Israel was the first to benefit: the almost automatic alliance between Arab countries and the Communist bloc had ceased to exist.

With the Gulf War, the USA regained its leadership in the Western world and ensured its presence in the Middle East. The Gulf War began between Wednesday 1 August and Thursday 2 August 1990, with the invasion of Kuwait by Iraqi military forces. After the invasion, the Kuwaiti royal family fled, and the country remained under Iraqi control. The Iraqi dictator Saddam Hussein had accused the Gulf States and especially Kuwait of producing higher quantities of oil than those agreed to by OPEC. This oversupply, in his opinion, contributed to the reduction in oil prices and hence the decline in revenues from exports. At the same time, the Iraqi economy was crushed by high indebtedness, and the reduction in oil prices made it more difficult to service the high debt repayments and import essential goods. Indeed, starting in 1989, the rich Gulf States had disregarded the agreements signed by OPEC members, helping to wipe out the potential benefits of an expanding cycle of oil consumption. After the minimum of $10 per barrel reached in 1986, the price of oil had risen gradually, reaching in January 1990 peaks of $22–23 per barrel. It therefore seemed that a new cycle of price expansion was starting, fuelled by a new increase in oil demand. It was not to be: the oil price fell to $13–14 a barrel in June 1990 (Clô, 2000). This new fall in prices was due to conflict among OPEC countries, a clash caused by those countries which wanted to increase their production even if it depressed world oil prices. The production surplus was especially due to Kuwait, the United Arab Emirates and Saudi Arabia. This oversupply led to a reduction in prices that benefited the consumer and oil companies.

Despite Hussein's accusations and threats to former Gulf allies, their behaviour did not change. Oil prices began to rise only after the markets began to perceive the risks inherent in Hussein's threats, in view of the already decided invasion of Kuwait. The USA, the international community and part of the

Arab world opposed the invasion. As a reaction, a multinational military force was sent to the Middle East. It was a coalition army, sponsored by the UN, but constituted and directed mostly by the USA, which represented the largest component. In addition, the UN proclaimed an embargo against Iraq. This was a boomerang against the Iraqi economy, which relied mainly on oil revenues. To compensate for the loss of production from Iraq and Kuwait, other OPEC members increased their production. The world was experiencing the sixth post-war oil crisis, and OPEC was living one of its worst moments. However, the international oil market was able to react efficiently to the "supply gap". At the end of the crisis, oil prices were exactly as they were at the start, around $18–20 per barrel. Thus the two peaks recorded during the conflict had been completely reabsorbed. Compared to the crises of the 1970s, the reaction of the industrialized countries was different. Above all, the panic effect was not triggered because Western countries had ample stocks. There was closer cooperation between industrialized countries within the IEA and in the meantime the weight of oil in the energy balance had already begun to decrease (Clô, 2000), thanks to the use of other sources of energy.

Although military intervention in the Persian Gulf undoubtedly resulted from the desire to restore Kuwait's sovereignty and respect for the rules of international law, it cannot be denied that oil played more than a secondary role (Clô, 2000). The definitive annexation of Kuwait would have allowed Iraq to almost double its directly controlled oil reserves, while its overwhelming military superiority in the region would have allowed it, in the absence of any intervention, to exercise a leading role also over Saudi Arabia, which was aware of the risk and therefore accepted to take on huge financial burdens for the *Desert Shield* military operation. On 6 August 1990, the Security Council of the UN approved a resolution that stopped all economic transactions to and from Iraq and occupied Kuwait. On 7 August, Saudi Arabia authorized the use of its territory and bases for the US army and its allies to prevent further advances of Saddam Hussein's troops. International markets suffered a strong backlash from the embargo on oil imports from Iraq and Kuwait, resulting in an increase in oil prices, which hit mainly Japan and Europe, and to a lesser extent the USA.

Although the USA was the first to react to Iraq's aggression and complete the liberation of Kuwait, it actually suffered less from the Gulf crisis than other economies. The success of the embargo against Iraq was also due to the contribution of the Arab countries, which increased their production, exceeding the ceiling previously established within the OPEC Geneva Conference. The war ended on 28 February 1991 with the announcement of US President George Bush to cease fire, after Kuwait had been liberated and Iraq had accepted all the resolutions of the UN. Despite the war, and economic and human losses, Hussein remained firmly in power. Moreover, the defeat of Iraq, the strengthening of more moderate Arab regimes and the return of the American presence to the Middle East gave the impression that the situation had stabilized. However, subsequent events showed that this was pure illusion. On 20 March 2003, the Second Gulf War began, with the invasion of Iraq by the US-led coalition. The stated goal was the end of Saddam Hussein's regime, accused of wanting to use weapons of mass destruction and form links with Islamic terrorism. Its opponents claimed that it was an "oil war".

The former *Rais* Saddam, in power in Iraq since 1979, was captured on 13 December 2003 by US troops and then executed in 2006. The weapons of mass destruction with which President George W. Bush had motivated the invasion of Iraq were never found, and the aim of making the world a safer place after the terrorist attacks on 11 September 2001 with the overthrow of the Saddam regime proved to be a tragic illusion. Especially in the Mediterranean region, geopolitical tensions concerning the security of energy supplies and especially of oil, which had increased again from the end of 2010 with the explosion of the Arab Spring, spread to Middle Eastern and North African countries: Tunisia, Syria, Algeria, Egypt, Libya, Jordan, Yemen, Iraq and Morocco. The war in Libya and the end of Gaddafi's regime in 2011 reduced Libyan oil exports almost to zero. The violent clashes that took place in Tripoli at the end of August and beginning of September 2018 demonstrated once again the precariousness of the internal equilibrium on which the Government of National Accord wished by the UN rests (Varvelli, 2018). Libya's instability continues to have major repercussions in terms of human lives, migrant flows and energy supplies. Also Syria, which plays a strategic role especially for energy transit, after about eight years from the beginning of the war is still being devastated by bombing. The war in Syria began in 2011 as a revolt against the authoritarian regime of Assad. On the night of 13–14 April, the USA, France and the UK bombed three military targets linked to Syrian President Bashar al-Assad. Over the years, however, the war in Syria has turned into something much bigger, since external powers have intervened – especially the USA, Russia, Iran and Turkey. Moreover, the great powers have not yet found a solution, despite the incalculable number of children, women and men who have been killed during the fighting and in spite of the great migratory wave of survivors seeking refuge in more secure European countries. Both in the case of Libya and Syria, there are many political, religious and economic interests at stake, among which is the control of the oil and natural gas fields and territories crossed by oil and gas pipelines. In a few words: the present and future security of energy supply.

## NOTES

1. We refer to Herbert Hoover, Jr, the former President's son, who US President Eisenhower appointed special representative of the USA in October 1953 with the aim of reaching a new oil agreement with Iran (Kapstein, 1990).
2. Qatar terminated its membership in January 2018; Indonesia suspended its participation in January 2009, reactivated it in January 2016, but suspended it again in November 2016. Ecuador suspended its membership in December 1992, but reactivated it in October 2007; Gabon opted out in 1995 and rejoined in July 2016. For further details see https://www.opec.org/opec_web/en/about_us/24.htm (last accessed 10 February 2019).

# 4. Energy transitions and energy efficiency

## 1. ENERGY EFFICIENCY AND PAST ENERGY TRANSITIONS

To better understand future energy transitions of emerging economies, it is important to identify and analyse the factors that have driven the past energy transition in those countries that now have developed economies. In recent years, policymakers and academics have considered fuelwood (biomass) as a "modern" energy source. They probably forget or do not know that firewood and charcoal were the main energy sources in Europe and the USA at least until the first half of the 1800s. Thus, from a historical perspective, "fossil fuels" are considered the modern energy sources, while fuelwood the traditional energy source. Once again in a historical perspective, it may be said that during the last two centuries, the energy system has changed, shifting from traditional energy sources to fossil fuels. For this reason, it is important to define the main phases of change in the energy system. Energy transitions from traditional energy sources to fossil fuels are considered a necessary, although not sufficient, condition to overcome the limits to economic growth (Bartoletto, 2012a). In particular, economic historians have stressed the role of technological progress, showing that the use of modern energy sources, first of all coal, was possible only thanks to technological progress, which makes it possible to use new energy converters able to harness new modern energy sources (Fouquet, 2008; Bartoletto, 2013a; Kander et al., 2013). Indeed, the existence of coal was well known in previous centuries, but only after the invention of the steam engine was it possible to use coal as fuel, and thus increase production capacity.

As stressed by Fouquet (2008, 2013), energy consumption derives from the demand for energy services, and the latter has changed over time. Hence the transition from traditional to modern energy sources was very complex and involved different sectors and services at different times. One of the main determinants of the energy transition was population increase. In Europe, the sharp increase in population between 1800 and 1850 accelerated the substitution of firewood with coal, since the former was insufficient to satisfy the demands of a growing population and, remaining within an energy system based only on traditional energy sources, would have triggered a new Malthusian trap (Bartoletto, 2012a). Another major factor that caused an increasing use of fossil fuels was the growth of urbanization. Indeed, the growth of cities was at the same time both a cause and effect initially of the transition to coal, and later oil and natural gas, other than nuclear, to produce electricity. A city can survive and grow only if it can harness energy from the surrounding environment. Especially from the first half of the nineteenth century, with the spread of the consumption of coal, cities began to change and grow rapidly (Allen, 2009; Bartoletto, 2004, 2007). This is another crucial factor to consider for the Mediterranean region, because at present the emerging economies, especially in North Africa, are experiencing rapid population growth and urbanization, which have contributed, and continue to do so, to increasing consumption of fossil fuels.

The transition from traditional to modern energy sources happened at different times in different countries. In England, at the beginning of the nineteenth century, coal represented more than 90% of total energy consumption (Warde, 2007). Yet England represented an exception, because in the rest of Europe, as in the USA, the use of coal during the nineteenth century was very modest. The transition from firewood and charcoal to coal was a slow process, and the two energy sources coexisted for a long time. Until the beginning of the twentieth century, much of Western Europe still depended on traditional energy sources. In France, Italy, Spain and Portugal, fossil fuels became the dominant energy source from World War II onwards. Another important aspect is that, over time, the world has experienced several energy transitions: the first was from firewood and other traditional energy sources to coal; the second from coal to oil; the third transition from oil to natural gas and nuclear. In recent years the energy markets have been revolutionized once again, this time by non-conventional energy sources such as shale gas and shale oil.

Energy transitions have entailed a change in energy efficiency, as shown by "energy productivity", which is the ratio between gross domestic product (GDP) and energy consumption. Thus energy productivity is the inverse of energy intensity. Bartoletto (2013a) estimated energy productivity in Europe from 1800 to 2000, showing that during the early stages, new machinery fuelled by coal had lower productivity than machinery fuelled by traditional energy sources such as fuelwood. As a consequence, in the first phases, GDP growth stemmed mainly from an increase in fuel input, rather than an increase in energy productivity. This may be ascertained from a decomposition of per capita GDP, as the result of per capita energy consumption ($E/P$) multiplied by the productivity of energy ($Y/E$). From this perspective, we can evaluate the importance of the growth rate of per capita energy consumption, and energy productivity in the growth of per capita GDP. For Western European countries, it is evident that from 1850 until 1913, a considerable decline took place in energy productivity (Table 4.1), especially during the sub-period 1850–1890. Energy input contributed more than energy productivity to GDP growth.

*Table 4.1 Rates of growth of per capita GDP (y), per capita energy consumption (e) and productivity of energy (π) in Western Europe (1820–2001)*

| Years | *y* | *e* | π |
|---|---|---|---|
| 1820–1850 | 0.42 | 0.32 | 0.10 |
| 1850–1870 | 0.50 | 0.65 | −0.16 |
| 1870–1890 | 0.51 | 0.70 | −0.19 |
| 1890–1913 | 0.63 | 0.65 | −0.02 |
| 1913–1921 | −0.58 | −1.18 | 0.60 |
| 1921–1931 | 0.86 | 0.47 | 0.39 |
| 1931–1941 | 0.96 | 0.18 | 0.78 |
| 1941–1951 | 0.19 | 0.32 | −0.13 |
| 1951–1961 | 1.76 | 0.78 | 0.98 |
| 1961–1971 | 1.51 | 1.59 | −0.08 |
| 1971–1981 | 0.98 | 0.20 | 0.78 |
| 1981–1991 | 0.84 | 0.11 | 0.73 |
| 1991–2001 | 0.70 | 0.37 | 0.33 |

*Source:* Bartoletto (2013a).

From World War I onwards, per capita GDP growth was fuelled both by an increase in energy productivity and energy consumption. The growth rate of energy productivity was much higher than that of per capita energy consumption during the periods 1913–1921, 1931–1941 and 1971–1991, peaking during the decade 1951–1961. Between 1900 and 1960, the growth rate of energy productivity in Europe was noteworthy. Energy input contributed less than energy productivity in per capita product growth. Acceleration of modern growth resulted from major innovations such as the internal combustion engine, electricity and the transition from coal to oil. Between the two World Wars there was a marked acceleration of technological progress. New machines were introduced, and even though many of the innovations originated in the previous periods, they were improved and became widespread during the wars. Technological progress led to a huge increase in economic efficiency, reducing costs and increasing production flexibility. Electric motors revolutionized low-cost power for industry and agriculture and stimulated the mechanization of production. Cars, trucks and tractors transformed transport costs and modes for both people and goods.

During the 1960s energy productivity diminished because oil, which in the meantime had become the main energy source, was abundant and cheap. After the severe oil crisis of the 1970s, the large increase in energy prices resulted in more efficient use and a decline in energy consumption in the industrial sector. Instability of oil-producing countries and high energy prices, together with environmental concerns, drove Europe to proceed in the direction of energy saving.

The trend of energy efficiency in a historical perspective is an important aspect that should be considered when studying present and future energy transitions. Each energy transition has a cost, generally associated with lower energy efficiency of the new energy sources. History tells us that it is a sort of necessary step, but thanks to technological progress, this constraint can be overcome. After the Kyoto Protocol, the new transition from fossil fuels to the production of electricity from renewables fuelled an important debate for several reasons, such as the lower efficiency of renewable energy with respect to fossil fuels. In the meantime, thanks to technological innovation, efficiency of renewables has significantly improved. Thus, the study of energy transitions in a long-term perspective is important because the lessons of the past show us the obstacles of the changes in the energy systems and help us better to predict the time and mode of future energy transitions in emerging countries. In a recent article, Agovino et al. (2019) analysed energy intensity to GDP from 1800 to 2000 for several European countries, distinguishing between traditional energy intensity and modern energy intensity. They demonstrated, through empirical and narrative analysis, that traditional and modern energy intensity should be considered separately, and that traditional energy sources are dominant when GDP levels are low, while modern energy carriers become dominant when GDP levels are high. In particular, traditional energy intensity shows a hyperbolic relationship with GDP per capita; conversely, modern energy intensity shows a reverse U-shaped relationship with per capita GDP.

Quality of energy is an important factor in energy efficiency because not all energy sources have the same economic productivity (Stern, 2010). Individual energy carriers have different properties and require specific technologies. Oil is of higher quality than coal or firewood because it can do work that solid fuels cannot, such as run a combustion engine (Reynolds, 1994). Electricity is the energy carrier with the highest quality, hence its widespread use. For example, coal cannot be used to power a computer directly, while electricity can (Stern, 2010).

To analyse energy intensity, it is important to consider the structure of the economy and the energy carriers consumed in each sector. The decline of energy intensity as countries' GDP increases is due either to technological or structural change or some combination of the two (Stern, 2011; Voigt et al., 2014). In the last two decades, the share of renewable energy sources on total consumption has significantly increased, especially in the most advanced countries of the EU. Moreover, the growth of industry and the transport sector in emerging countries is driving strong growth of fossil fuel consumption and $CO_2$ emissions. This particularly holds for the Mediterranean region, especially for the growth of energy demand from North African and Middle Eastern countries. For this reason, a historical perspective on the relation between energy intensity and different GDP groupings of advanced European countries of the Mediterranean region could provide policymakers with important suggestions. In the early stages of economic development, the share of industry on total output increases, causing a large increase in energy consumption in the industrial sector. In the latter stages of economic development, the share of services on GDP increases, while that of industry decreases. Since services are less energy-intensive than industry, the so-called dematerialization of economies contributes to a reduction in energy intensity (Medlock and Soligo, 2001). As countries shift from low to high GDP, modern energy intensity increases. By contrast, total energy intensity and traditional energy intensity decrease when shifting from low to high GDP (Agovino et al., 2019). This is the case of Italy and Spain, since as they shift from low to high GDP, modern energy intensity increases (Figures 4.1 and 4.2). By contrast, total energy intensity and traditional energy intensity decrease when GDP switches from low to high.

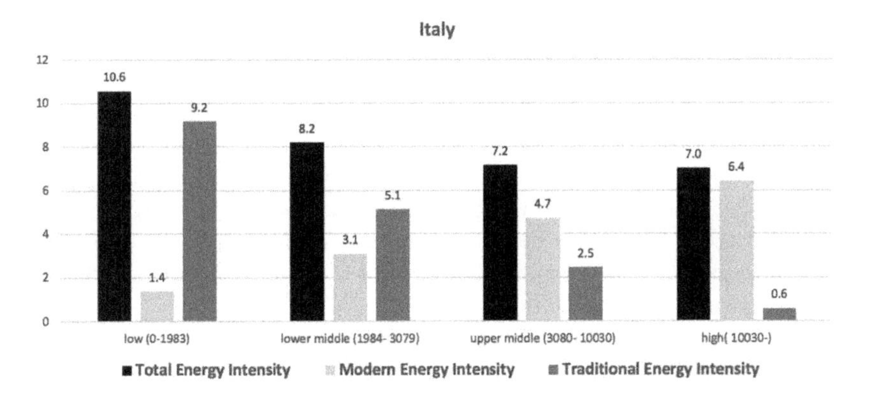


*Figure 4.1* *Italy: energy intensity by GDP grouping (1861–2000)*

*Note:* GDP per capita is in 1990 international dollars, energy intensity is in MJ/1990 int. $.
*Source:* Agovino et al. (2019).

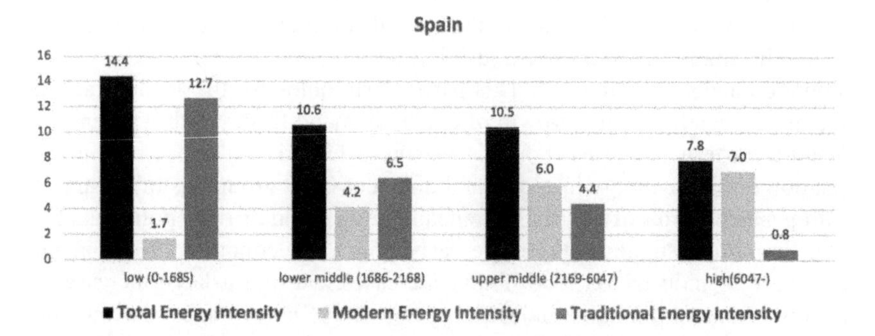


*Figure 4.2* *Spain: energy intensity by GDP grouping (1861–2000)*

*Note:* GDP per capita is in 1990 international dollars, energy intensity is in MJ/1990 int. $.
*Source:* Agovino et al. (2019).

In Italy, when GDP per capita was low, agriculture represented the main sector of the economy and total energy intensity corresponded almost to traditional energy intensity. Starting from the early 1900s to the eve of World War II (lower-middle GDP), the share of industry on GDP increased significantly together with coal consumption and modern energy intensity, while traditional energy intensity declined. From the end of World War II to the eve of the oil crises (upper-middle GDP), industry became the most important sector and GDP per capita increased significantly. Modern energy intensity continued to grow, especially due to the rapid growth of oil consumption. From the 1970s to 2000 (high GDP), per capita GDP increased together with modern energy intensity. Fossil fuel consumption continued to grow, but after the 1970s' oil shocks, the growth rates of natural gas consumption rapidly increased, while that of oil diminished. The value added of industry peaked in the early 1980s, and then declined, while value-added services increased, becoming the most important sector in GDP. Energy consumption in the industrial sector fell sharply from 1973–1974 and in 2001 it was overtaken by the transport and service sector (Bartoletto, 2005).

## 2. ENERGY INTENSITY

Energy intensity is an important macroeconomic indicator that expresses the relationship between the energy consumption of a country and its GDP. It is often interpreted as a measure of the efficiency of a nation's economic system and as an indicator of sustainable development. However, interpreting energy intensity as a synthetic indicator of the efficiency of an economic system may lead to erroneous conclusions being drawn. This emerges clearly when considering the Mediterranean area, which includes 25 countries which differ greatly in economic conditions, levels of development and energy consumption.

Energy intensity expresses the consumption of energy required to produce a unit of income and is calculated through a quotient that has, in the numerator, energy consumption, expressed in units of energy (in our case in toe), and GDP in the denominator. To be more precise, primary energy intensity is calculated as the ratio between total primary energy consumption (TPES) and GDP, while the final energy intensity is obtained by dividing total final energy consumption (TFC) by GDP. There are numerous factors that can lead to a change in energy intensity, such as the variation in individual sectoral intensities, which is a mainly technological factor, as it describes the change in the unit energy requirement for the production of a given good or homogeneous group of goods; the change in the composition of the national product, or the change in the weight of the individual sectors in the economy of the country; the change in the amount of energy directly required for end use by households, such as for heating or for private transport.

The reciprocal of energy intensity is energy productivity, that is, the relationship between GDP and energy consumption. Energy productivity is a measure of the technical efficiency of conversion and hence of factor productivity. As a first approximation, a decline over time of energy intensity beyond a certain level of income could be interpreted in terms of greater efficiency of the economic system. What increases or decreases the energy intensity quotient is the change in energy consumption (numerator) compared to GDP. If the numerator (energy consumption) increases less than the denominator (GDP), then the energy intensity decreases.

Over the years, energy intensity (TPES/GDP) has decreased worldwide, from 0.22 in 1971 to 0.126 in 2016.[1] However, there are important differences between the different areas. While in the Organisation for Economic Co-operation and Development (OECD) countries the levels of energy intensity are below the world average, in non-OECD countries energy intensity remains above the world average despite having considerably decreased in recent decades.

By analysing the trend of energy intensity during a shorter or longer period, we can thus evaluate the extent to which decoupling of energy consumption and economic growth has occurred. Yet to understand the reasons that influence the trend, we need to undertake other investigations since energy intensity is an indicator that cannot fully capture the complexity of the energy–economy nexus. We need only consider China and the USA, which have reduced their respective energy intensities, but nevertheless represent the countries with the highest energy consumption and $CO_2$ emissions.

China is the world's most populous country and consumes the largest quantities of energy, followed by the USA. China's GDP is estimated to have grown

9% in 2008, while average growth rates were 10% between 2000 and 2008. Although the recent economic and financial crisis has significantly reduced GDP growth rates, China's energy demand remains very high. From being a net exporter of oil in the early 1990s, in 2006 China became the world's third largest importer of crude oil, rising to second place in 2014 and first place by 2018.[2] Consumption of natural gas and related imports have also grown rapidly over the last few years, and it is currently the second largest net importer of natural gas worldwide. In addition, China is both the most important producer and a net importer of coal, which is much more polluting than oil and natural gas.[3] As a result, while China's energy intensity decreased from 0.955 in 1971 to 0.152 in 2016,[4] thanks above all to the strong growth of GDP, energy consumption grew rapidly, as did the relative emissions of $CO_2$.

The change in the energy intensity of a given country over time reflects the different nature of the link between the growth in energy consumption and the underlying level of economic development. Therefore the energy intensity curve shows a trend that varies according to the state of economic growth: the curve grows rapidly during industrialization until it reaches its peak at the stage of industrial maturity; subsequently energy intensity tends to decrease as income increases and therefore the corresponding curve decreases thanks to the growth of the role of high value-added services. According to a very common interpretation among economists, energy consumption does not increase proportionally to GDP. The relationship between energy and GDP (energy intensity) shows what resembles an inverted U-shaped curve, that is, a curve that grows in the early stages of industrialization and then declines in the post-industrial phase.

Such arguments, however, do not take into account the traditional energy consumed prior to the introduction of fossil fuels. It is reasonable to believe that the long-term trend of the curve would be different if it were to include traditional sources of energy, such as wood, food for humans and working animals, water and wind (Agovino et al., 2019). A fundamental role has been played by technological progress, thanks to which further economic growth becomes possible with decreasing energy intensity. This leads to greater optimism regarding the future scenario: the depletion of fossil energy sources can, at least in part, be compensated by increased technological efficiency.

A decisive role in the performance of energy intensity is played by the price of fuels. When the latter rises, energy intensity tends to decrease, leading to a more efficient use of energy. Indeed, reduction in energy intensity means energy savings per unit of product and is caused, first of all, by the desire to save money in the production process in order to maintain or increase profits, as demonstrated by past experience. During the 1950s and 1960s, when energy was plentiful and cheap, energy intensity was high. Following the oil crises of the 1970s and rising oil prices, energy intensity decreased globally. To conclude, it may be stated with certainty that energy intensity, albeit an important indicator, is not sufficient to explain the complexity of the link between energy, economy and the environment. In particular, from an environmental point of view, everything depends on the amount of energy consumed and the mix of fuels used. Although two countries may have the same energy intensity or show the same trend over time, there may be important environmental differences between them. In other words, even without a significant reduction in energy intensity, the environmental impact of energy consumption can be reduced, for example as a result of a change in the fuel mix. This is because, with the variation of the fuels used and hence the technologies used, the efficiency rates according to which the primary energy is converted into useful energy vary. There are therefore many variables that can affect the trend of energy intensity and which do not emerge from simple consideration of the value of the indicator.

## 3. ENERGY EFFICIENCY IN NORTH AFRICAN COUNTRIES

Improving energy efficiency is fundamental for reducing energy consumption, air pollution (Evans et al., 2013) and improving energy security. Improving energy efficiency also produces benefits at macro-level for other reasons. For example, the reduction of energy costs leads not only to greater energy access and increasing energy productivity but also to the improvement of private and public budgets.

One of the indicators used for estimating energy efficiency is energy intensity, which measures the energy required to produce a unit of economic value. However, low energy intensity does not necessarily mean high energy efficiency, as stated above.[5] To evaluate energy efficiency in Mediterranean countries, we consider total final energy consumption and energy intensity by sector.

The fuel mix plays a key role in energy efficiency. Hence the transition from one energy system to another based on a different energy basket is one of the main determinants of changes in energy efficiency. In previous sections we considered total primary consumption by the fuel used. Now we focus on final energy consumption by fuel, to better evaluate the trend of energy efficiency.

Final energy intensity depends on the level of final energy consumption and GDP. Final energy consumption, in turn, depends on the structure of the economy, which is reflected in the demand for energy services such as transport, lighting, cooking, heating, industrial goods and so on. Another important determinant of energy efficiency is the energy mix. In 2016, Africa's crude oil production continued to decline, led by Nigeria and Libya, while increasing in Algeria. By contrast, the production and consumption of biofuels (mainly fuelwood) across Africa is much higher than the world average due to the presence of large forests, agroindustry, agriculture and a large rural population. Indeed, in Africa, the use of fuelwood for cooking is very high. A low GDP per capita, combined with the extensive use of wood and charcoal with its low energy efficiency, leads energy intensity in Africa to be higher than the world average.[6] The energy balance of North African countries is different: here the energy mix is dominated by fossil fuels, while fuelwood consumption is below the world average. Indeed, in 2016, the share of biofuel consumption on total final consumption was respectively 54% in Africa, 3% in North Africa, and 11% worldwide.[7]

It is important to recall that only part of firewood consumption is accounted for in the official statistics. There is a substantial amount of firewood consumption for domestic use which is not included.

Allowing for this important limit of official statistics, within the North African region, Morocco and Tunisia have the highest percentages of charcoal and firewood consumption for cooking, albeit well below the average of African countries.

Figure 4.3 shows the share of biofuels on total final consumption, and its evolution during the last fifty years. Interestingly, at the beginning of the 1970s, biofuels represented more than 12% of total final consumption in North Africa. Yet if the single countries are analysed, the share is much higher, and we have a picture of the energy transition of North African countries from fuelwood to fossil fuels. In 1971, in Libya, biofuels still represented 12.5% of total final consumption. Their share had rapidly decreased after oil production and exports began during the 1960s. Indeed, during the 1980s, biofuels declined to less than 3% of TFC of Libya and currently they represent about 1.5%.

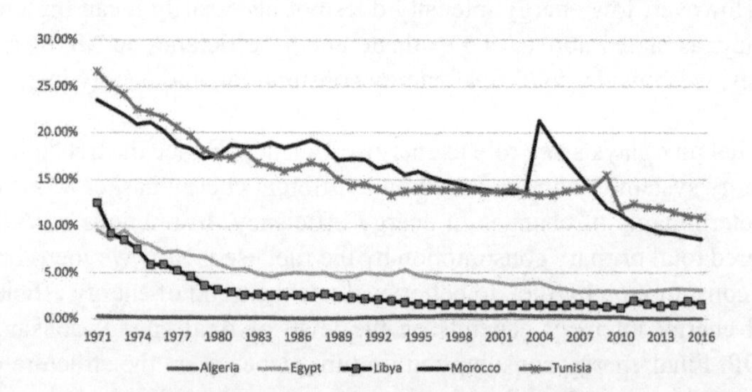


*Figure 4.3* *Share of biofuels on total final consumption in North Africa, 1971–2016 (%)*

*Source:* Our calculations based on IEA, World Energy Balances, data extracted on 15 June 2019.

From Figure 4.3 it is evident that in all countries biofuel consumption is decreasing, especially fuelwood, while the contribution of fossil fuels is increasing. Tunisia and Morocco have always been the countries with the highest percentages of biofuels in total final consumption. In the same period, the share in Egypt and Libya has decreased. This means that in Egypt and Libya the transition from fuelwood to fossil fuels started before. In the case of Libya we find a watershed represented by oil discoveries during the 1950s.

The share of biofuels on total final consumption in Algeria has always been close to zero during the period investigated, also at the beginning of 1970s. This different composition of the energy basket is reflected in the pattern of final energy intensity (Bartoletto, 2013b). Algeria, whose energy balance at least from 1970 was founded exclusively on fossil fuels, had lower levels of final energy intensity than the other countries more based on biofuels. The situation has radically changed in very recent years, when final energy intensity of Algeria rapidly increased (Table 4.2, Figure 4.4).

*Table 4.2 Final energy intensity ranking of North African countries (selected years)*

| Countries 1972 | | Countries 2005 | | Countries 2010 | | Countries 2016 | |
|---|---|---|---|---|---|---|---|
| Morocco | 0.07 | Tunisia | 0.08 | Libya | 0.07 | Libya | 0.22 |
| Egypt | 0.07 | Egypt | 0.07 | Tunisia | 0.07 | Algeria | 0.07 |
| Tunisia | 0.07 | Libya | 0.07 | Egypt | 0.07 | Tunisia | 0.07 |
| Algeria | 0.02 | Morocco | 0.07 | Morocco | 0.06 | Egypt | 0.06 |
| Libya | 0.01 | Algeria | 0.05 | Algeria | 0.06 | Morocco | 0.06 |

*Source:* IEA, World Indicators, data extracted on 3 June 2019.

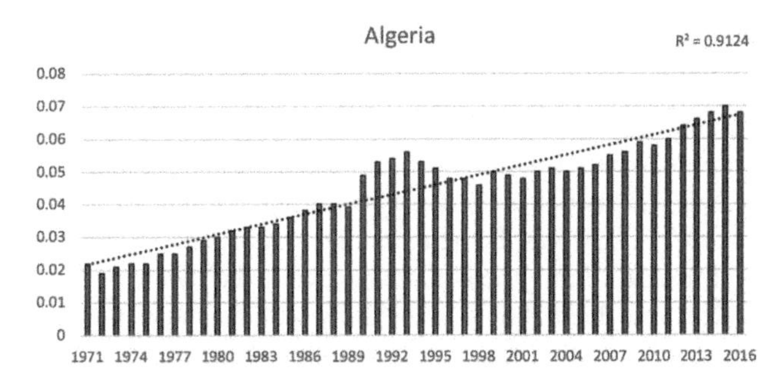


*Figure 4.4 Final energy intensity in Algeria, 1971–2016 (toe per thousand 2010 USD PPP)*

*Source:* Our estimation based on IEA, World Indicators, data extracted on 3 June 2019.

The rapid growth of total final energy intensity is due to the growth of final energy intensity of all sectors. Transport is the sector with the highest energy intensity, followed at a considerable distance by residential and industry. From 1971 to 1989, the levels of energy intensity of the residential and industrial sectors were quite similar. From 1990 onward, the gap between the two sectors increased. The industrial sector continues to be underdeveloped since the main revenue source of the country still consists of oil and gas exports. In the industrial sector and the residential sector, the main energy source is natural gas, while the transport sector is based completely on oil products (Figure 4.5).

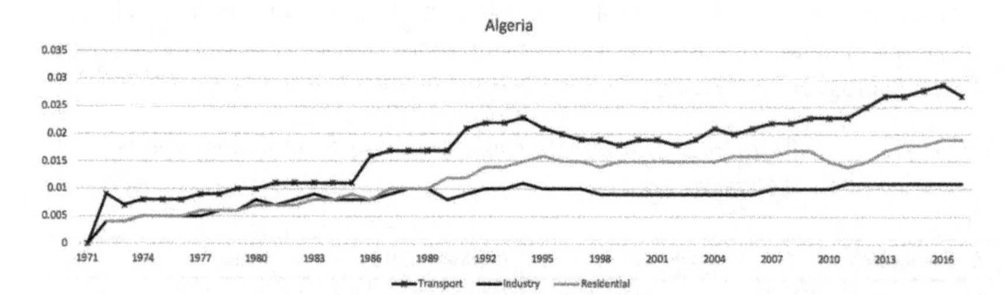


*Figure 4.5* *Algeria: energy intensity by sector, 1971–2016 (toe per thousand 2010 USD PPP)*

*Source:* IEA, World Indicators, data extracted on 3 June 2019.

If the final energy intensity of single North African countries is compared, there is a clear process of convergence. At the beginning of the 1970s the gap between countries was wide. Subsequently, the growth of final energy intensity in Algeria and Libya, and the contraction in Egypt and Tunisia, led to a decrease in the divide within the North African region.

In Egypt, final energy intensity has followed a wavelike pattern, with phases of recovery and reduction (Figure 4.6). In more recent years, after the peak in 2007, final energy intensity has declined. As to single sectors of the economy, final intensity of industry was the main component of total final energy intensity from 1971 to 2010. From then on, it was overtaken by the transport sector (Figure 4.7). This was due not only to the growth of the latter, but especially to decreasing energy intensity throughout the period examined. The main energy sources are oil and natural gas. By contrast, the residential sector remained almost stable during the period examined.

*Figure 4.6* *Final energy intensity in Egypt, 1971–2016 (toe per thousand 2010 USD PPP)*

*Source:* IEA, World Indicators, data extracted on 3 June 2019.

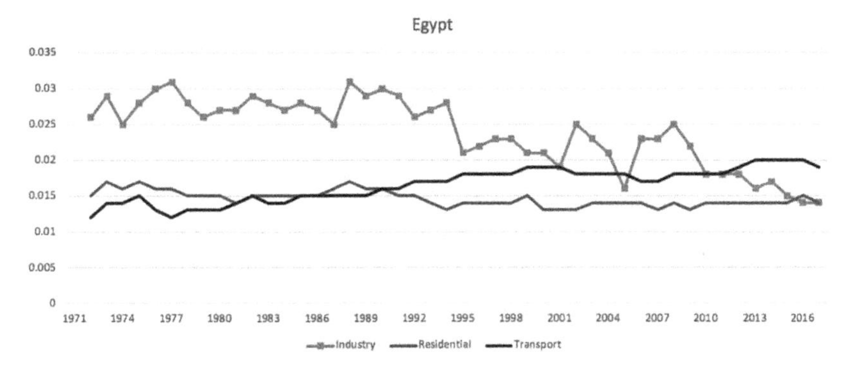


*Figure 4.7* *Egypt: energy intensity by sector, 1971–2016 (toe per thousand 2010 USD PPP)*

*Source:* IEA, World Indicators, data extracted on 3 June 2019.

Energy self-sufficiency of Egypt has drastically declined during the last fifty years.

The case of Tunisia, where there has been a significant reduction of final energy intensity, is different. With respect to other North African countries, until 2005 Tunisia presented the highest levels of energy intensity (Table 4.2). Subsequently, final energy intensity declined (Figure 4.8). This was due to the reduction of energy intensity in the industrial and residential sectors. While the latter has shown a declining trend during the period examined, energy intensity in the transport sector increased, and in recent years has become the sector with the highest energy intensity (Figure 4.9). Another interesting aspect of Tunisia is the high level of energy intensity of the residential sector from 1971 to 2016. First, the level of residential energy intensity is higher than in other North African countries, and only in 2013 was it overtaken by Algeria. Second, it is also higher than other more advanced countries, such as Italy and France.

A higher level of residential energy intensity stems mainly from a higher share than other countries of biofuel consumption. In Tunisia, in 2016, biofuel consumption represented more than 27% of total final consumption in the residential sector.

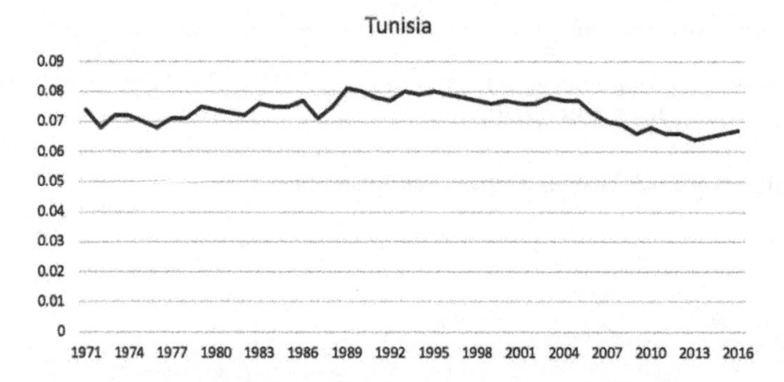


*Figure 4.8* *Final energy intensity in Tunisia, 1971–2016 (toe per thousand 2010 USD PPP)*

*Source:* IEA, World Indicators, data extracted on 3 June 2019.

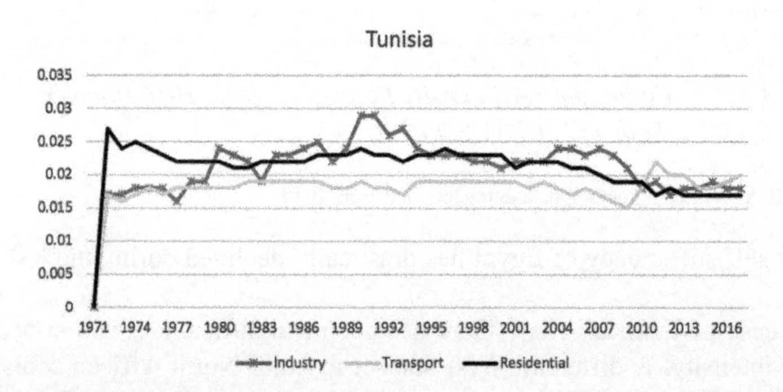


*Figure 4.9* *Tunisia: energy intensity by sector, 1971–2016 (toe per thousand 2010 USD PPP)*

*Source:* IEA, World Indicators, data extracted on 3 June 2019.

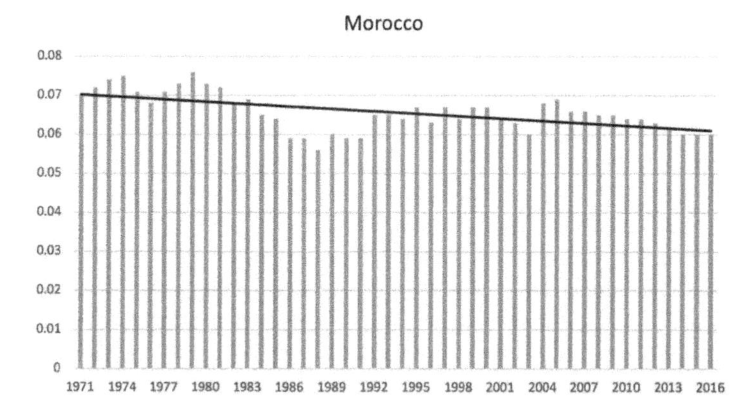


*Figure 4.10* *Final energy intensity in Morocco, 1971–2016 (toe per thousand 2010 USD PPP)*

*Source:* IEA, World Indicators, data extracted on 3 June 2019.

In the case of Morocco, from inspection of the graph in Figure 4.10 it is clear that phases of reduction and growth have taken place during the last fifty years, even though a declining trend is evident.

As for decomposition by sector, the role of industry, transport and residential sectors has changed over time. Industry was the most energy-intensive at the beginning of the period analysed. Like other countries, energy intensity in the industrial sector has declined, while that of the transport sector has rapidly increased, especially since the beginning of the 1990s, and the sector is now more energy-intensive. The residential sector overtook the industrial sector at the beginning of the 2000s (Figure 4.11).

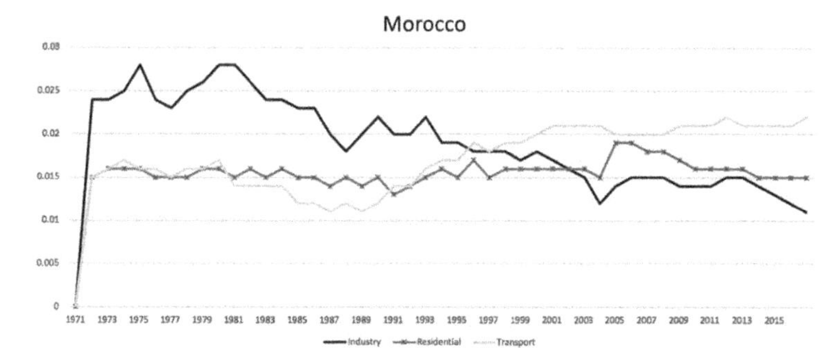


*Figure 4.11* *Morocco: energy intensity by sector, 1971–2016 (toe per thousand 2010 USD PPP)*

*Source:* IEA, World Indicators, data extracted on 3 June 2019.

Almost 60% of total final consumption is satisfied through oil products, which are the main energy source also in the residential sector. In the latter, biofuels in 2016 represented about 14%. Morocco is experiencing rapid growth, which led to a rapid increase in energy consumption, an increase in oil in the energy balance and a reduction of renewables in total final consumption. This is translated into strong growth of energy imports and energy dependency.

In Libya, final energy intensity has experienced a growing trend (Figure 4.12). At the beginning of the 2000s, final energy intensity had decreased significantly, but from 2011 the pattern has been affected by the war and GDP reduction.[8] In 2011, when the Gaddafi regime was deposed, final energy intensity drastically declined. There was an upsurge in 2012, in 2013 another crash, and then another upsurge.

In Libya the most energy-intensive sector has always been that of transport. While in other North African countries, transport overtook industry only in a subsequent phase, in Libya, the transport sector has always registered the highest levels of energy intensity (Figure 4.13). In 2016, the transport sector represented about 71% of total final consumption and all the demand of the transport sector was satisfied through oil products. Final energy consumption in industry is only 6% of the total, while the residential sector accounts for about 16–17%. The use of natural gas is mainly concentrated in the latter.

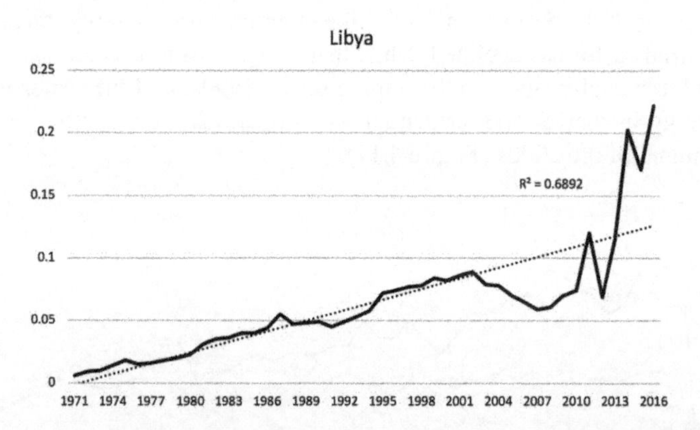


*Figure 4.12* *Final energy intensity in Libya, 1971–2016 (toe per thousand 2010 USD PPP)*

*Source:* IEA, World Indicators, data extracted on 3 June 2019.

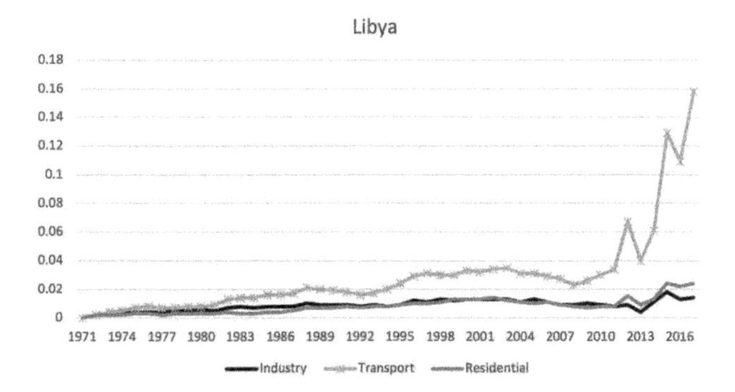


*Figure 4.13 Libya: energy intensity by sector, 1971–2016 (toe per thousand 2010 USD PPP)*

*Source:* IEA, World Indicators, data extracted on 3 June 2019.

# NOTES

1. IEA, World Indicators, data extracted on 25 June 2019. Values are expressed in toe per thousand 2010 USD.
2. For 2018 data: IEA, *Key World Energy Statistics*, 2018. For previous years: IEA, *Key World Energy Statistics*, 2016, and Bartoletto, 2016a.
3. IEA, *Key World Energy Statistics*, 2018. Data refer to 2017, and are provisional.
4. IEA, World Indicators, data extracted at www.iea.org. on 25 June 2019. We refer to People's Republic of China.
5. IEA, *Energy Efficiency Indicators*, 2016.
6. IEA, *World Energy Balances*, 2018.
7. It is important to recall that this share includes not only biofuels, but also waste.
8. IEA, *World Energy Balances*, 2018. "Due to new information on oil and electricity becoming available from 2006, breaks in time series may occur between 2005 and 2006".

# 5. Renewables and $CO_2$ emissions

## 1. RENEWABLE ENERGY SOURCES

An extensive debate over the years has developed among economists, environmentalists, policymakers and others as to what extent environmental and resource constraints will limit economic growth. The various perspectives have been grouped into three categories in Hepburn and Bowen (2013). The first is that environmental factors do not limit economic growth, thanks to technological progress over time, in accordance with standard neoclassical and endogenous growth models (Lomborg, 2001). The second perspective refers to the concept of "environmental drag" introduced by Nordhaus (1992), in response to a wide debate that started after the publication in 1972 of the work entitled *The Limits to Growth* by Meadows et al. Constraints to long-term economic growth, imposed either by environmental concerns or natural resource limitations, were recognized by Nordhaus, but he concluded with regard to past growth trends that "resources have been but a small drag on the growth to date and that technological change has overwhelmed the small drag".[1] As for future trends, he predicted that "environmental and resource constraints on economic growth should be only modest over the half century".[2] In his analysis, technological change is the key factor to outpace population growth and resource exhaustion.

The third, most pessimistic perspective is that economic growth cannot continue indefinitely because environmental limitations are significant enough to prevent sustained growth in consumption and production.

The study of energy transitions (Fouquet, 2008; Bartoletto, 2012a) has shown that the development of technology is one of the main factors of economic growth. Since stopping economic growth would be counterproductive, it is wrong to consider economic growth as a problem. Recessions have slowed or cancelled cleaner production programmes. We need more sustainable growth, we need a transition to a low-carbon economy, and not zero growth. Renewable energy sources – wood, windmills, water mills, the first hydroelectric power plants – dominated world energy balances until the end of the nineteenth century, before being overtaken by fossil fuels in the countries that were undergoing industrialization, but remaining dominant in less developed contexts.

With the emergence of the climate issue as a priority in the political agenda of governments, a new phase opened up in the history of renewables. Despite higher generation costs, and the limits of discontinuity, randomness and low productivity of renewables that have not been solved by advances in technology, renewable energy has developed mainly thanks to public policy support. Starting from the 1970s, the green movements gained increasing political weight in many European countries such as Germany, Italy, Belgium and Finland, and thanks to their political pressure the protection of the environment became enshrined in the policies of the European Community. With the implementation of the Kyoto protocol, an expansionary cycle of investments in new renewables was initiated, which ensured high profitability for many years thanks to the combined certainty of quantities and production subsidies. With the new policies in support of renewables it was no mere replacement of primary sources, of fossil fuels with renewables, but the establishment of a new technological paradigm. While until that moment, when fossil fuel consumption was at its peak, attention had focused on the greater efficiency of plants, the maximization of utilization rates and the minimization of systemic costs, now instead with the consumption of renewables, more small size capital-intensive plants were used, with low programmability and lower utilization rates. The more renewable energy penetrated, the more the electricity systems had to acquire flexibility due to the unpredictability of supply (Clô, 2017). At the same time, the liberalization processes of the electricity markets favoured the multiplication of new operators and new generation plants on a region-wide basis, creating new problems in the electricity markets. Since the start of the twenty-first century, there has been extraordinary growth in renewable investments worldwide. Renewable electric power has greatly increased and the highest growth has focused on wind and solar, thanks to incentives and reduction in the cost of renewables. However the contribution in terms of generation has been lower as a consequence of low use and their intermittence. The growth of investments in renewables has been impressive especially in Europe, reaching a peak in 2011, and then slowing for several reasons, such as the slump in fossil fuel prices, but especially due to the economic crisis that led to the cutting of economic incentives for renewables, which has reduced industrial investments. Thus in Europe and the USA the great expansive phase is running out of steam (Bloomberg, 2018). In 2017, China was the leading location for renewable investments, with an exceptional solar boom. In contrast, in the US renewable investment was down 6% in 2017, and Europe suffered a major decline of 36% (Bloomberg, 2018, p. 11). Europe's share of global investment in renewable energy reduced to 15% in 2017, the lowest recorded since 2004 (Bloomberg, 2018, p. 20).In some developing countries the trend is different, due to higher incentives by foreign companies. Morocco, for example, is developing renewable energy policies. This country has no fossil fuel reserves, but it has plenty of sunshine, and developing solar energy production means reducing its dependence on energy supplies from abroad. The crucial aspect of the issue is to find a way to sustain the development of renewable technologies within a difficult market. One of the main arguments to support the development of renewable energy is the politically increasingly complex issue of energy security, that is, the problem of dependence on unstable areas. A new energy transition is required but past experience shows that much time is needed, that each country has its own specific issues, and that times and modes of energy transition change from one country to another for several reasons, such as their endowment of energy sources, economic development and climate conditions. With existing technologies, it is impossible to imagine that renewable energy sources will completely replace fossil fuels in the coming decades.

To avoid facile public policies concerning future energy transitions it is important that policymakers know the economic, political, industrial and social obstacles in the path of previous energy transitions. The process of replacing one source of energy with another depends not only on the evolution of technology but also on the outcome of decisions taken by public and private stakeholders.

Throughout history, energy has played an important role in economic growth, because it has been a boost when new energy sources and technologies were used, but it has been responsible for slowing growth rates during periods of perceived scarcity (Fouquet, 2013). The transition to a low-carbon energy system could be a boost for the economy or, in turn, could create new limits to economic growth. Several studies have focused on the transition to a renewable energy system, showing that diminishing fossil fuel reserves are not the main incentive. Rather, the transition would appear due to preferences for renewable energy or to the latter becoming cheaper than fossil fuels (Fouquet, 2013). It is demonstrated that a complete transition to a low-carbon economy is unlikely if renewable energy provides more costly energy services than fossil fuels (Fouquet, 2010).

## 2. EURO-MEDITERRANEAN COOPERATION FOR ENERGY AND CLIMATE CHANGE

The Mediterranean region is considered a major hotspot of climate change, whose adverse effects could be both substantial and numerous (Benoit and Comeau, 2005). For example, the neighbouring Sahara Desert has expanded significantly over the twentieth century, with estimates varying from 11% to 18% (Thomas and Nigam, 2018). Climate change across sub-Saharan Africa would have a disastrous impact on North Africa, with severe repercussions for Southern Europe, including a further increase in migration flows.

As the Mediterranean region is currently experiencing a sharp increase in population, it is expected that in the coming decades there will be a strong rise in the demand for energy, especially in southern Mediterranean countries. What will contribute to rising energy demand and climate change emissions is not only the increase in population, but also the development of the industrial sector and the growth of tourism. Thus a major effort is required to accelerate the transition to a low-carbon economy. Several steps have been taken by the European Union (EU) to promote dialogue and cooperation with the southern and eastern shores of the Mediterranean. A fundamental step was the Euro–Mediterranean Conference of Foreign Affairs Ministers, held in Barcelona on 27 and 28 November 1995, which defined the political, economic and social framework of the relations between the EU and those of the Mediterranean area. With the Barcelona Declaration, the signatories undertook to achieve three main objectives, namely to identify a Euro–Mediterranean area of peace and stability based on the fundamental principles that include respect for human rights and democracy, to create an area of shared prosperity through economic-financial alliance, and the progressive liberalization of trade between the EU and its partners, and among the Mediterranean countries themselves. The Barcelona Conference also took a fundamental step forward in energy policy: the pivotal role of the energy sector in the Euro–Mediterranean economic partnership was acknowledged and it was decided to strengthen cooperation and intensify dialogue in the field of energy policy. With the Barcelona Declaration, the signatories decided "to create the appropriate framework conditions for investments and the activities of energy companies, cooperating in creating the conditions enabling such companies to extend energy networks and promote link-ups".[3] With a view to creating appropriate conditions for investment and activities by energy companies, it was established that cooperation would focus on oil and gas exploration, refining, transportation, distribution, and regional and trans-regional trade; coal production and handling; generation and transmission of power and interconnection and development of networks; energy efficiency; new and renewable sources of energy; energy-related environmental issues; and development of joint research programmes. Thus within the Barcelona process, an important role was given to energy, but not only to renewable production, but also to exploration, production and trade in fossil fuels.

Europe is faced with the problem of high energy dependency and security of its energy supply. It therefore encourages the exploration and production of fossil fuels in the less developed partner countries of the Southern Mediterranean area, with the aim not only to promote economic development of these countries, but also to secure its energy supply.

The Euro–Mediterranean partnership has resulted in bilateral agreements between the EU and each Mediterranean partner country, and in the launch of the MEDA programme in 1996 to provide financial and technical assistance to countries in the southern Mediterranean. However, only about 26% of the MEDA funds committed for the period 1995–1999 was actually spent (Pace, 2003). A review was therefore necessary as the MEDA programme was slowed down by procedural complexity and rigidity. The EU therefore adopted a new regulation in 2000 called MEDA II,[4] launched with the aim of making Community financing swifter and more effective. Within the MEDA programmes, the construction of new energy infrastructures was to be financed.

The Barcelona Conference and the launch of the MEDA programmes were followed by a void in political attention towards the Mediterranean: the focus of European policies was on the enlargement of the EU across Central and Eastern Europe. On 25 October 2005, in accordance with the Council Decision of 17 October 2005, the Energy Community Treaty was signed.[5] The Energy Community Treaty provides for the creation of an integrated energy market in South-East Europe which will create a stable regulatory and market framework capable of attracting investment in gas networks, power generation and networks, such that all Parties have access to stable and continuous energy supplies that are essential for economic development and social stability.[6] Euro–Mediterranean cooperation, especially the Barcelona process, was relaunched in 2008 with the creation of the Union for the Mediterranean (UfM) which was agreed on 13 July 2008 during the French Presidency of the EU. The UfM is an intergovernmental institution bringing together the 28 EU Member States and 15 countries from the southern and eastern shores of the Mediterranean to promote dialogue and cooperation. All the countries of the Mediterranean region take part in the UfM, except for Syria, which suspended its membership on 1 December 2011, and Libya, which has observer status.

Moreover, current upheavals across the Middle East and North Africa, which spread after the "Arab Spring" in 2011, pose the question of whether these structures and tools are suitable to tackle a global picture which has rapidly changed. Moreover, climate change remains a major security issue for several reasons. The most important potential source of renewable energy for the Euro-Mediterranean region is the sun. To exploit this potential the DESERTEC project was launched,[7] since enormous amounts of energy are delivered every day by the sun to deserts. The DESERTEC project seeks to expand energy cooperation between Europe and MENA countries, namely Middle Eastern and North African countries (Bloomfield et al., 2011). The potential of solar energy is also recognized by the Mediterranean Solar Plan (MSP), which is one of the six key initiatives of the Union for the Mediterranean. However, economic crises and political instability in MENA countries have slowed down the implementation of these projects. At the same time, although the slogan of UfM is "in energy, the goal is to create an integrated Mediterranean energy market, and to promote renewables and energy efficiency", several new projects are being developed to increase not only renewable production, but also fossil fuel production in EU Neighbourhood-South.[8] Algeria, for example, is the third biggest supplier of gas to the EU, and the EU is the biggest importer of Algerian gas. Europe relies on Algerian gas to secure its supplies and Algeria relies on the European market to secure demand. This strong interdependence in the energy sector has led to the establishment of an EU–Algeria Strategic Partnership on Energy. It covers cooperation on natural gas, renewable energy, energy efficiency and energy market integration. Launched in 2015, it is now fully operational.[9]

On 23 April 2018 the EU and Egypt signed a new Memorandum of Understanding in Cairo on a Strategic Energy Partnership, updating their 2008 Memorandum of Understanding on Energy. This new memorandum focuses on six areas, stemming from EU–Egypt partnership priorities: (1) further assistance to the oil and gas sector; (2) electricity sector reforms; (3) development of an energy hub (oil and gas hub); (4) further assistance with joint measures and projects in the field of renewable energy; (5) additional support on energy efficiency strategies, policies and measures across various sectors; (6) enhancement of cooperation between Egypt and the EU in technological, scientific and industrial areas across the energy sector.[10] The strategy of the EU is clear: it helps neighbouring countries to develop renewable energy but at the same time implement oil and gas production to secure its own energy supply.

Although Southern and Eastern Mediterranean countries have set ambitious goals for renewables, the actual implementation of national plans is still quite slow. In recent years, important joint research programmes have led to major discoveries, such as the discovery of a new gas field in Egypt, thanks to the cooperation with the Italian energy company ENI. This suggests that, in the next few decades, renewable energy sources are highly unlikely to replace fossil fuels completely. Egypt is one of the countries in the Mediterranean area that has recorded the highest rates of growth of $CO_2$ emissions, like other North African countries. Also Israel abandoned its climate policy, following the discovery of huge natural gas reserves (Michaels and Tal, 2015).

Although investment in electricity from renewable sources is slowing (Bloomberg, 2018), in the EU the contribution of renewable energy sources remains higher than the world average. From a regulatory point of view, for the implementation of renewable energies there is the Directive 2009/28/EC on the promotion of the use of energy from renewable sources.[11] This Directive, amending and repealing the previous Directives 2001/77/EC and 2003/30/EC, creates a common framework for the use of renewable energy in the EU in order to reduce greenhouse gas emissions and promote cleaner transport. To this end, it sets targets for all EU countries in order to bring the share of energy from renewable energy sources to 20% of all EU energy and 10% for the transport sector by 2020. This Directive is one of the 20-20-20 objectives of the EU Energy and Climate Change Package. The other two objectives are: to reduce greenhouse gas emissions by 20% compared to 1990 levels and to improve energy efficiency by 20% by 2020. Despite several contradictions in environmental policies, the EU Energy and Climate Change Package has led the share of renewable energy consumption to double over time, and $CO_2$ emissions to be significantly reduced.

There are several mechanisms that can help achieve the above objectives. The guarantee of achieving the desired level of emissions can be obtained through a system of tradable permits. In practice, the legislator sets the goal, distributes it among the subjects, through the assignment of securities representing the emissions, and allows them to exchange such securities. The final allocation of the bonds between actors and the consequent determination of the unit cost of the emission is thus entrusted to the market; the latter corresponds to the point of intersection of the supply and demand curves of securities. A system of trading emission such as that described is termed a "cap and trade" system: the legislator sets the desired quantity of emissions (cap) and distributes to the participants in the scheme a number of tradable permits (allowances) corresponding to the above quantity. The most important examples of the cap and trade system are the EU emission trading scheme (EU ETS), and the emission trading system introduced by the Kyoto Protocol. At the beginning of the 1990s most of the scientific community agreed on the possible negative effects of the increase in atmospheric concentration of greenhouse gases. In 1992, the UN Framework Convention on Climate Change was signed in Rio de Janeiro (UNFCCC). The convention established the principle of common but differentiated responsibility, by virtue of the different contribution in terms of emissions of industrialized countries and developing countries. The industrialized countries were formally committed to adopting emission reduction policies compared to 1990 levels. The convention provides for Conferences of the Parties to be held annually for the discussion of the formulation and evaluation of international climate policy.

The third conference of the parties, held in 1997, ended with the enactment of the Kyoto protocol, an agreement through which the industrialized countries are committed to achieving reduction targets defined by the protocol itself. Article 3 of the protocol sets a target for the removal of various climate-changing gases, which include $CO_2$, methane and nitrous oxide. The Kyoto protocol came into force only in 2005, as it was necessary to exceed the minimum thresholds set for ratification, in terms of the number of adherents and the share of emissions represented by these parts. The protocol was ratified by the EU, Japan, Canada and Russia, but not by the USA. It was also recently abandoned by Canada.

To enable the objectives to be achieved for each party obliged, the Kyoto protocol envisaged three market mechanisms, through which the countries subject to reduction obligations can use securities that represent abatement measures carried out in third countries. With the emission trading mechanism, the Kyoto protocol introduced a cap and trade system between the states committed to achieving the emission reduction targets. Each objective translates into the availability of emission certificates, which can be freely exchanged between the participants. In this way, countries that are able to reduce their emissions beyond the limits set by the protocol can sell excess securities to countries that exceed emission limits. With the "clean development mechanism", in addition to the bonds that represent the caps of each participating country, states can use credits deriving from abatement projects carried out by public or private entities in developing countries. These are called certified emission reduction (CER) credits. With the "joint implementation mechanism", states that have a reduction obligation can also carry out joint reduction projects in the territory of a country subject to the obligation, from which they originate credits that can be used to fulfil the obligations established by the protocol. Several European countries, among which Italy, launched initiatives in the countries of North Africa and in the Balkans, which generated emission credits or carbon credits through the Clean Development and Joint Implementation mechanisms. In October 2014, EU leaders agreed on new climate and energy targets for 2030, including: a reduction of (at least) 40% of greenhouse gas emissions compared to 1990 levels; a minimum share of 27% of energy from renewable sources; a minimum improvement of 27% in energy efficiency. For energy efficiency and renewable energy, the original target of at least 27% was revised upwards in 2018. The new targets are 32% for renewable sources and 32.5% for energy efficiency.[12] In the long term, more incisive global cuts will be needed to avoid dangerous climate change.

The Paris Agreement established a new market mechanism to replace the "clean development mechanism and the joint implementation" after 2020.[13] The EU is committed to reducing emissions by 80–95% from 1990 levels by 2050, provided that developed countries participate in the collective effort. One of the backbones of the EU's measures to limit climate change continues to be the exchange of emission quotas. Moreover, the EU has recently recognized that the surplus of allowances is largely due to the economic crisis (which reduced emissions more than anticipated) and high imports of international credits.[14] This has led to lower carbon prices and hence a weaker incentive to reduce emissions. Thus the system needs to be improved. The new "European Green Deal" does not currently envisage continuing use of international credits in EU ETS after 2020.[15]

## 3. RENEWABLE ENERGY CONSUMPTION IN MEDITERRANEAN COUNTRIES

In the Mediterranean region, renewable sources account for 11% of total primary energy supply, with a contribution which has grown over the years due to increased efforts to exploit the high potential that the region has in the sector of renewable energy. Although major efforts have been made to develop solar energy, at present about half of the total renewable energy consumption is represented by biofuels, whose consumption has rapidly increased in recent years, and now represents about 6% of total energy consumption in the Mediterranean region.

Biofuels and waste comprise solid biofuels, liquid biofuels, biogases, industrial waste and municipal waste. Biofuels are defined as any plant matter used directly as fuel or converted into fuel (e.g. charcoal) or electricity and/or heat. This includes wood and plant waste (including wood waste and crops used for energy production). While biomass occupies a prominent position, it is difficult to have an exact estimate of the consumption of this source, especially in the case of firewood and other agricultural by-products, which are regularly collected and used as fuels to heat households and for cooking, especially in developing countries. The problem is that this type of consumption escapes official statistics and is likely to be underestimated. Even in more developed countries the same problem can be found. In the case of Italy, for example, a survey carried out by ENEA (2001) showed that the consumption of firewood was far higher than that resulting from the national energy balance, where only commercialized wood is counted.

Solar and wind energy combined represent 18% of total renewable consumption, and only 2% of total primary supply. This share is expected to increase in the years to come, especially in Morocco, which has made large investments in solar energy, and Egypt. After biofuels, the main renewable energy source is hydroelectricity, which supplies 23% of total renewable consumption, but only 2% of total primary supply. This marginal role is due to the presence of vast drought-affected areas in the Mediterranean region. By contrast Italy and the Balkans are well-endowed with water resources, unlike North Africa where 97% of Egypt and 90% of Libya consist of desert. Thus there are major differences between the countries belonging to the EU and the Southern and Eastern Mediterranean Countries (SEMCs). About 60% of hydroelectricity is consumed in the countries of the Latin area, especially in Italy, where hydroelectricity production is quite high, and in France where, however, the role of hydroelectricity has been eclipsed by nuclear power.

In the Fourth Report of the Intergovernmental Panel on Climate Change (IPCC), the Mediterranean is considered one of the regions of the world most exposed to the effects of global warming, which will cause a worsening of the problem of water scarcity that already affects many countries on its southern and eastern shores.

The SEMCs are faced with rapid population growth, combined with low income and rapid urbanization. This translates into growing demand for energy services and related infrastructures. The SEMCs are facing various environmental problems: desertification, growing coastal urbanization, air and water pollution, waste disposal, oil pollution of coastal waters, and increasing emissions of greenhouse gases. In this context and in an international context of growing environmental concerns, most SEMCs have stated that they wish to pursue a model of sustainable growth. Through the Mediterranean Renewable Energy Programme (MEDREP), efforts have been made to encourage the use of renewable energy in the Mediterranean basin. The main objectives are to increase the share of renewable energy resources in order to reduce $CO_2$ emissions and therefore mitigate climate change, and also to equip rural populations with modern energy services. MEDREP is a policy instrument that was set up specifically for the Mediterranean region. It was launched by the United Nations (UN), and several countries and international organizations have joined. The specific objectives of MEDREP are to deliver electricity to isolated rural populations, based on village-scale mini-grids; accelerate integration of renewable energy in the national electricity grids; and implement the use of renewable energy, mainly solar, in the building sector. The above objectives will be achieved with the implementation of innovative pilot projects in the renewable energy sector.[16] For example, through MEDREP funds, Egypt has electrified some rural villages by using renewable energy. Apart from this, other projects have been carried out by the EU. In spite of this, in the Mediterranean region there is a certain slowness of investments despite a repeatedly expressed political desire to favour renewables (MEDENER-OME, 2018).

Although all countries, within the framework of the Paris Climate Conference, have made commitments to reduce climate-changing emissions, the pace and magnitude of efforts varies from country to country. The war in Libya and Syria, the unresolved dispute between Israel and Palestine, and more generally the climate of instability in the Mediterranean region have contributed to the slowdown in investments. Rapid population growth has led to a rise in energy demand which has increased the energy deficit. Forecasts show that population and energy demand will continue to increase in future years. Solutions are urgently required to improve energy efficiency and energy distribution systems, reduce losses and develop technologies that lead to an increase in renewable production.

In Turkey, the explosive economic growth that occurred in the mid-1990s had significant repercussions on energy consumption and the environment.

Currently, hydroelectricity production is increasing, and it accounts for about 24.5% of total electricity production, even though about 67% is generated by thermoelectric power plants, mainly fuelled by coal and gas. Coal-fired generation accounted for 33% of total electricity production in Turkey in 2016.[17] However, over the last few years hydroelectricity has increased significantly and a further significant increase is expected in the coming years, as Turkey is taking part in an ambitious project involving the construction of new hydroelectric plants along the Tigris and Euphrates river basins.[18]

In Egypt, on the other hand, hydroelectricity production has decreased over the years, despite the increase in energy demand due to population growth, improvement in living standards, the construction of new industrial areas and expansion of the commercial and residential sectors. Whereas the dam on the Nile River supplied a good part of Egypt during the 1980s, currently 70% of electricity production derives from thermoelectric plants powered by natural gas, while hydroelectricity accounts for just 9%, largely deriving from the basin of the Aswan Dam.[19] From the last report on renewable energy by the Mediterranean Association of National Agencies for energy management (MEDENER),[20] and OME, during the period from 2010 to 2015 solar and wind energy recorded the highest growth rates among renewable energies in Turkey, Egypt and Morocco (MEDENER-OME, 2018).

The production of geothermal energy is scarce and is essentially concentrated in two countries: Italy and Turkey. Italy represents one of the countries with the highest levels of geothermal production worldwide. Production is concentrated in Larderello in Tuscany, where there is one of the first geothermal plants created in the world.

In Table 5.1 we show the energy balance of North African countries. Although they have high potential for renewable energy, especially solar, at present the common characteristic is that fossil fuels represent more than 90% of total consumption, except for Tunisia, where fossil fuels amount to 88.6%. In Algeria, renewable energy represents only 0.06% of total consumption. Algeria has important oil and gas reserves, and it is the most important exporting country within the Mediterranean region. Starting from 2011, it has implemented a renewable energy plan, subsequently reviewed in May 2015. This plan calls for solar and wind power to be developed. According to this new plan, by 2030, renewables should represent 27% of electricity production and 37% of total installed capacity. Moreover, at present, almost 100% of total primary energy supply is represented by fossil fuels. The main energy source is natural gas, with a share of about 64%, followed by oil, with about 35% of the total.

*Table 5.1* *North Africa: total primary energy supply by fuel in 2016 (per cent)*

| Countries | Oil | Coal | Natural Gas | Renewables |
|---|---|---|---|---|
| Algeria | 35.4 | | 64.4 | 0.06 |
| Egypt | 43.9 | 0.40 | 51.97 | 3.68 |
| Libya | 71.1 | | 27.6 | 1.01 |
| Morocco | 63.3 | 22.5 | 5.4 | 8.6 |
| Tunisia | 41.2 | | 47.4 | 10.7 |

*Source:* Our calculations based on IEA, World Energy Balances, data extracted on 20 November 2018.

Egypt has a population of 93.8 million and its energy demand is growing fast. The objective of Egypt is to reach 20% from renewable energy of installed capacity by 2022. At present, renewable energy supplies only 3.7% of total consumption (see Table 5.2), while fossil fuels supply about 96%. The main energy source is natural gas, with a share of about 52%. Egypt is an important gas producer, and in recent years its reserves have expanded significantly. Oil is the second energy source in order of importance, with a share of 44%.

In Tunisia, the development of renewable energy is linked to the Tunisian solar plan, which started in 2009 and was amended in 2015. The plan envisages that by 2030 solar energy will reach 30% of electricity production. At present, Tunisia is the North African country with the highest percentage of renewable energy, corresponding to 10.7% of total primary supply in 2016. The country has high biomass consumption, which represents about 90% of renewable energy consumption. The percentage would be much higher if the calculation were to consider the quantities of wood consumption that are omitted from official statistics. Solar energy represents only 8% of total renewable energy consumption, and only 0.9% of total primary energy supply. Moreover, national plans are banking on this energy source.

In Morocco renewable energy supplies about 8% of total consumption, while fossil fuels account for more than 90%. Morocco has major plans to expand renewable production, especially solar and wind, but it is far from easy to forecast whether the share of renewable energy will increase because energy demand is increasing rapidly at the same time. One of the main objectives undertaken by Morocco is to diversify energy sources to secure its energy supply.

*Table 5.2 North Africa: share of renewables on total primary energy supply, 1971–2016 (per cent)*

| North Africa | 1971 | 1990 | 2000 | 2016 |
|---|---|---|---|---|
| Algeria | 1 | 0 | 0 | 0 |
| Egypt | 14 | 6 | 6 | 4 |
| Libya | 6 | 1 | 1 | 1 |
| Morocco | 26 | 14 | 12 | 8 |
| Tunisia | 25 | 13 | 13 | 11 |

*Source*: IEA, World Indicators, data extracted on 2 January 2019.

On 7 April 2019, at the time of writing, conflict flared up again in Libya, causing hundreds of victims. The data presented herein refers to 2016, when fighting had ceased and political instability put a halt to economic recovery. In 2016, fossil fuels represented 99% of total consumption, with renewable energy supplying only 1%. There is no law to facilitate the promotion of renewable energy. On the institutional level there was the creation of the Ministry of Electricity and Renewable Energy. Libya's general electricity company (GECOL) is a vertically integrated company that has a monopoly on production, transmission and distribution (MEDENER-OME, 2018). In the meantime, the situation has deteriorated, and the fighting has broken out inside the country again.

In Middle Eastern countries, the share of renewable on total primary energy supply doesn't exceed 3%, because these countries, despite not having reserves of oil and gas, have focused on their imports to meet the growing domestic demand for energy. In Lebanon, the percentage has even reduced, from 9% in 1971 to 2% in 2016 (see Table 5.3).

*Table 5.3 Middle East: share of renewables on total primary energy supply, 1971–2016 (per cent)*

| Countries | 1971 | 1990 | 2000 | 2016 |
|---|---|---|---|---|
| Israel | 0 | 3 | 3 | 2 |
| Jordan | 0 | 2 | 1 | 3 |
| Lebanon | 9 | 7 | 4 | 2 |
| Syria | 0 | 2 | 2 | 1 |

*Source:* IEA, World Indicators, data extracted on 2 January 2019.

In Italy, the share of renewables on total primary consumption has almost tripled, from 6% to 17%; in Greece and Spain it has more than doubled, while in France the share of renewables has grown only by 4 percentage points

*Table 5.4* *EU countries: share of renewables on total primary energy supply, 1971–2016 (per cent)*

| Countries | 1971 | 1990 | 2000 | 2016 |
|---|---|---|---|---|
| Croatia | – | 13 | 19 | 24 |
| Cyprus | 2 | 0 | 2 | 7 |
| France | 9 | 7 | 6 | 10 |
| Greece | 8 | 5 | 5 | 12 |
| Italy | 6 | 4 | 6 | 17 |
| Malta | 0 | 0 | 0 | 4 |
| Portugal | 20 | 20 | 15 | 25 |
| Slovenia | – | 9 | 12 | 17 |
| Spain | 6 | 7 | 6 | 15 |

*Source:* IEA, World Indicators, data extracted on 2 January 2019.

because it has mainly banked on nuclear power (see Table 5.4). Albeit with differences between countries, it may be said that in all European countries the share of renewables has grown because the EU has set objectives to which individual countries must adapt. In particular, the "2030 climate & energy framework" established the binding goal of bringing the amount of energy consumption satisfied by renewable sources to at least 32% by 2030. The original target of at least 27% was revised upwards in 2018.[21]

## 4. $CO_2$ EMISSIONS IN MEDITERRANEAN COUNTRIES

Human activities associated with the burning of fossil fuels are contributing significantly to changing the world's climate. Reducing the negative impacts of climate change requires a drastic cut in greenhouse gas emissions, of which carbon dioxide ($CO_2$) is the most important source. World $CO_2$ emissions almost tripled from 1971 to 2016, rising from 13 945 to 32 314 Mt.[22]

Ecological disasters penalize low-income countries along coasts to a greater extent. For this reason the Mediterranean region is more exposed, especially on its southern shores where there is a higher level of poverty.

Climate change mitigation policies have focused on industrial and commercial sources, sometimes underestimating emissions from individuals and households (O'Garra, 2013). For this reason it is important to consider the moral issue of climate change. Any intervention useful for reducing climate-changing emissions has a monetary cost. A fundamental question is to determine who pays. In addition, while policies for progressive decarbonization are implemented at world level, at the same time non-conventional energy sources, such as shale oil and shale gas, have increased the supply of energy more than renewable energy sources have been able to do.

The Intergovernmental Panel on Climate Change (IPCC) (2018) has estimated that human activities have caused about 1.0°C of global warming above pre-industrial levels. If global warming continues to grow at the current rate, it is very likely that it will reach 1.5°C above pre-industrial levels between 2030 and 2052. As stressed by the IPCC (2018), "warming from anthropogenic emissions from the pre-industrial period to the present will persist for centuries to millennia and will continue to cause further long-term changes in the climate system, such as sea level rise, with associated impacts".

Land and ocean ecosystems have already changed as a consequence of global warming. Populations dependent on agricultural or coastal livelihoods, in other words much of the Mediterranean region, will experience a greater risk of negative consequences of global warming. If global warming of 1.5°C increases to 2°C, the negative consequences will be even worse on food availability, water supply, health, security and economic growth. Among the regions most exposed to this kind of risk, the IPCC lists the Mediterranean region.

The environmental question came to the attention of the international community in 1972 with the UN Conference on the Human Environment in Stockholm. Yet a real turning point in environmental policies came twenty years later, with the UN Conference on Environment and Development (UNCED), known as the Earth Summit, held in Rio de Janeiro from 3 to 14 June 1992, which was the first world conference of heads of state on the environment. It was an unprecedented event also in terms of media impact and consequent political and development choices.

For the first time, the atmosphere was recognized as a global public good: climate change in each country did not depend only on its own emissions but on the global concentration of carbon dioxide in the atmosphere. During the Earth Summit, the serious conflict between global dimension and national interests emerged, the historical responsibilities of the industrialized states for the enormous growth of the concentration in the atmosphere of $CO_2$ and the contrast between the states from which they originate and those (the poorest) that suffer the most from their effects.

Environmental degradation and growing economic inequality are "two key challenges" for policymakers for the next few decades (Chancel and Piketty, 2015). This aspect emerges more markedly for the Mediterranean region, where the richer countries of the EU lie opposite the poorer countries of North Africa. Within the Mediterranean region $CO_2$ emissions have doubled in the last fifty years, rising from 1049 Mt in 1971 to 2017 Mt in 2016.[23] Emissions are expected to rise significantly in the next few years, as an effect of growing demand for fossil fuels by North African and Middle Eastern countries.

However, if the rates of growth of $CO_2$ emissions are considered, then over time they have significantly declined (Figure 5.1). The ten-year growth rate of 29% for the period 1971–1981 declined to 18.4 during the subsequent period, 1982–1992. The oil crises of 1973–1974 and 1979–1981 led to a huge increase in oil prices and growing concern for security of energy supply. As a reaction, the demand for oil decreased, and was replaced by natural gas (which emits lower levels of $CO_2$), and nuclear (which emits no $CO_2$). From 1993 to 2003, the growth rate of $CO_2$ emissions increased again, reaching 26.9%. From then on, the $CO_2$ emissions growth rate decreased markedly in the Mediterranean region during the period 2004–2016, but only in part due to policies to abate $CO_2$ emissions, developed by different countries within the Kyoto Protocol, which became effective from 2005.

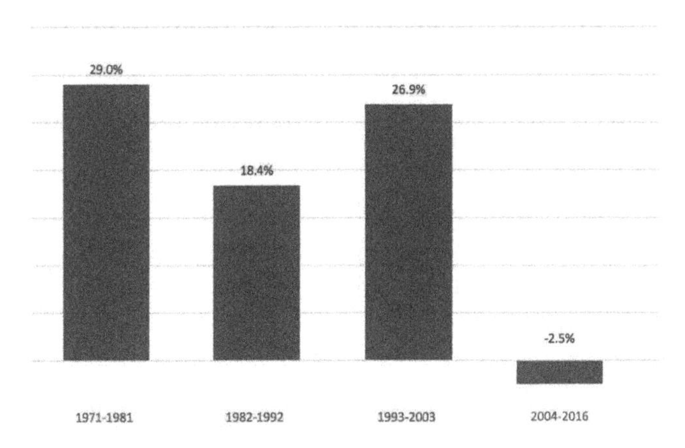


*Figure 5.1* *Growth rates of $CO_2$ emissions in the Mediterranean region, 1971–2016 (%)*

*Source:* Our calculations based on IEA, Indicators for $CO_2$ Emissions, data extracted on 21 December 2018.

In-depth analysis, developed at the level of single macroareas, allows us to calculate the rates of growth of $CO_2$ emissions within different macroareas of the Mediterranean region. From our calculations, it is evident that the reduction in $CO_2$ growth rates is only due to the strong reduction in European countries. By contrast, in other areas, the growth has been considerable. In North Africa, during the period 2004–2016, $CO_2$ emissions increased by 52% (Table 5.5). The growth would have been even higher without the Libyan crisis. The highest growth rate was registered in Algeria (73.8%), where oil and gas consumption increased rapidly over the same period. High rates of growth (above

50%) were also registered in Egypt and Morocco despite government efforts in both countries to increase production of renewable energy. In Tunisia, too, growth was sustained (32.6%). The most intense growth within the North African area occurred during the period 1971–1981, subsequently declining, albeit maintaining high growth rates, with a recovery occurring during the last period analysed.

*Table 5.5* *Growth rates of $CO_2$ emissions in North Africa (per cent)*

| Countries/Years | 1971–1981 | 1982–1992 | 1993–2003 | 2004–2016 |
|---|---|---|---|---|
| **North Africa** | 178.5 | 49.9 | 39.9 | 52.3 |
| Algeria | 273.3 | 44.0 | 27.0 | 73.8 |
| Egypt | 131.8 | 58.9 | 43.5 | 59.8 |
| Libya | 391.9 | 28.9 | 55.4 | 2.1 |
| Morocco | 110.6 | 52.4 | 45.2 | 51.1 |
| Tunisia | 118.9 | 68.4 | 33.6 | 32.6 |

*Source:* Our calculations based on IEA, Indicators for $CO_2$ Emissions, data extracted on 21 December 2018.

Morocco has long started reforms and plans for across-the-board development of industry, infrastructure, agriculture and energy. The North African country is focusing most of all on solar and wind to generate electricity for sustainable and lasting development, since it has no oil and gas reserves but a high potential for solar energy.

Despite these policies, $CO_2$ emissions increased by 51% from 2004 to 2016, because coal, which is the most polluting fossil fuel, continues to represent about 22.5% of total primary consumption. Coal imports have greatly

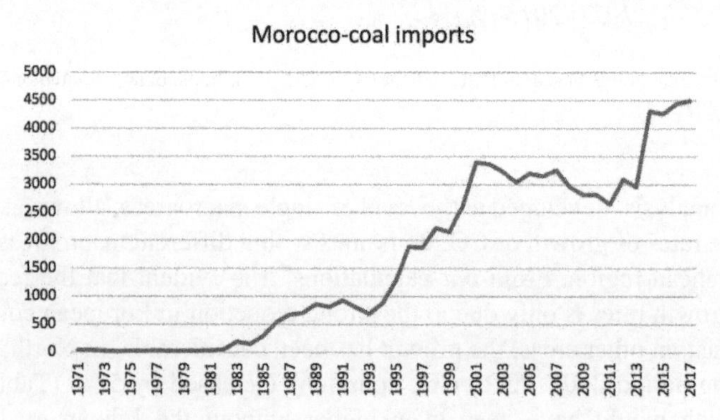


*Figure 5.2* *Morocco: coal imports from 1971 to 2017 (ktoe)*

*Source:* IEA, World Energy Balances, data extracted on 30 July 2019.

increased over time, and also in recent years, imports have continued to grow (Figure 5.2), although Morocco is pursuing policies to increase solar and wind energy. In addition, it is largely dependent on oil, which represents 63% of total primary consumption (Table 5.1).

In Libya, the growth rate of $CO_2$ emissions dropped from 55.4% from 1993–2003 to 2% in the period 2004–2016. This was due to the war and to the severe economic crisis.

A similar pattern may be observed in the Middle East. It could be said that the growth rate has decreased from 48% to 1%, but on reading Table 5.6, it may be seen that the latter result depends only on the considerable reduction by Syria, which has been devastated by war. Lebanon recorded a growth of 53%, and Jordan of 43%. Also for Middle Eastern countries, the most intense growth occurred between 1971 and 1981, with a ten-year growth rate of 79%, subsequently declining. From an in-depth analysis of Table 5.6, the pattern may be observed to differ within different countries. In Lebanon, the $CO_2$ emissions growth rate has increased over time. By contrast, in Israel and Jordan it has fallen.

*Table 5.6 Growth rates of $CO_2$ emissions in the Middle East (per cent)*

| Countries/Years | 1971–1981 | 1982–1992 | 1993–2003 | 2004–2016 |
|---|---|---|---|---|
| **Middle East** | 79 | 69 | 48 | 1 |
| Israel | 42 | 70 | 56 | 4 |
| Jordan | 279 | 77 | 37 | 43 |
| Lebanon | 37 | 47 | 50 | 53 |
| Syria | 157 | 70 | 41 | −39 |

*Source:* Our calculations based on IEA, Indicators for $CO_2$ Emissions, data extracted on 21 December 2018.

Moreover, in very recent years, policies in Israel have changed, and the national economic interest has prevailed over the climate change issue. After the discovery of a huge gas field, Israel abandoned objectives to reduce $CO_2$ emissions. Although Israel is located in the heart of the hydrocarbon-rich Middle East, it has no fossil fuel reserves, which is why the country has had a high level of energy dependency. Moreover, energy supplies are not provided by closer producers of the Middle East, because of "political animosity" (Bahgat, 2005). Indeed, Israel has had to import oil and natural gas from several countries as far afield as Mexico, Norway and Russia. The lack of energy resources, and the need to import energy from countries situated far away has been a heavy burden on the balance of payments for Israel. Hence, a key aspect of Israeli energy policy is to diversify its energy mix to avoid being overly dependent on one energy source.

Even though the larger share of energy demand is satisfied through oil, Israel is seeking to implement the use of natural gas and renewable energy. In 2009, there was the discovery of a huge natural gas reserve in Israel's territorial waters of the Mediterranean Sea (Michaels and Tal, 2015). Other fields continued to be discovered, like the Leviathan field in 2010. Thanks to these discoveries, Israel has reduced its level of energy dependency. Thanks also to gas production, the country has reduced its dependency on coal, which is the most polluting of fossil fuels. Thus the switch from oil and coal to natural gas lends an important contribution to reducing $CO_2$ emissions. Yet at the same time, focusing on natural gas and neglecting renewable energy will have a major impact on Israel's $CO_2$ emissions.

With respect to North Africa and Middle East countries, the pattern of $CO_2$ emissions for European countries that are part of the Mediterranean region is completely different. EU Member States have agreed on a new renewable energy target of at least 32% by 2030.[24] Indeed, in single European countries, $CO_2$ emissions have significantly fallen over time. Italy is one of the European countries which have most reduced $CO_2$ emissions. Thanks to the increase in renewable energy and natural gas in the energy mix, together with important progress made in energy efficiency, $CO_2$ emissions decreased by 28% from 2004 to 2016 (see Table 5.7). This is a major result, given that Italy has no nuclear plants. By contrast in France, which is the country at world level with the highest share of electricity produced by nuclear, the reduction in $CO_2$ emissions has been 21%, lower than Italy, which produces no electricity of nuclear origin. In Italy, the share of renewable energy on the total primary energy supply more than doubled from 2004 to the present day (from 7% to 17%). Also in Spain and Slovenia there has been a reduction in $CO_2$ emissions, even though both countries, like France, have nuclear in the energy mix.

*Table 5.7 Growth rates of $CO_2$ emissions in some European countries (per cent)*

| Countries/Years | 1971–1981 | 1982–1992 | 1993–2003 | 2004–2016 |
|---|---|---|---|---|
| Croatia | NA | NA | 36.5 | −18.9 |
| Cyprus | 47.1 | 80.8 | 42.9 | −8.7 |
| France | −4 | −8 | 8 | −21 |
| Greece | 78 | 58 | 30 | −33 |
| Italy | 20 | 13 | 16 | −28 |
| Malta | 57 | 69 | −7 | −46 |
| Portugal | 73 | 72 | 34 | −18 |
| Slovenia | NA | NA | 13 | −11 |
| Spain | 60 | 20 | 45 | −25 |

*Source:* Our calculations based on IEA, Indicators for $CO_2$ Emissions, data extracted on 21 December 2018.

Major progress has been made in Greece, where $CO_2$ emissions have decreased significantly. During the period examined, the growth rates have been declining: from 78% (1971–1981) to 58% (1982–1992) and 30% (1993–2003). During the last period (2004–2016), the rate of growth has even become negative mainly for three reasons: the severe economic crisis that led to a reduction in energy consumption; the increasing share of renewable energy sources, which has nearly doubled from 2004 to the present day; and the reduction of coal, which is the most polluting energy source. Also in Cyprus, the reduction of $CO_2$ emissions is due not only to the increase in renewable energy, but also to the reduction in coal consumption over time. This is an important aspect, because Greece and Cyprus, together with Turkey, used to be the most coal-based countries within the Mediterranean region. However, in Turkey, coal consumption continues to increase along with related $CO_2$ emissions. Indeed, in Turkey, $CO_2$ emissions increased by 63% during the period 2004–2016. Coal, particularly lignite, is Turkey's most abundant indigenous energy resource, and coal-fired power stations accounted for about 33% of electricity generation in 2016.[25]

In order to establish whether the gap between different countries of the Mediterranean region is narrowing, we calculated the Gini index on per capita $CO_2$ emissions for the period from 1971 to 2016. Figure 5.3 shows that the Gini index has decreased over time, falling from 0.43 in 1971 to 0.28 in 2016. The divide strongly narrowed from 1971 to 1986. Indeed, during this period, the variation range fell from 0.43 in 1971 to 0.29 in 1986. The gap then increased once again, reaching a peak of 0.34 in 1993, and 0.33 in 2003. From then on, it has continued to narrow.

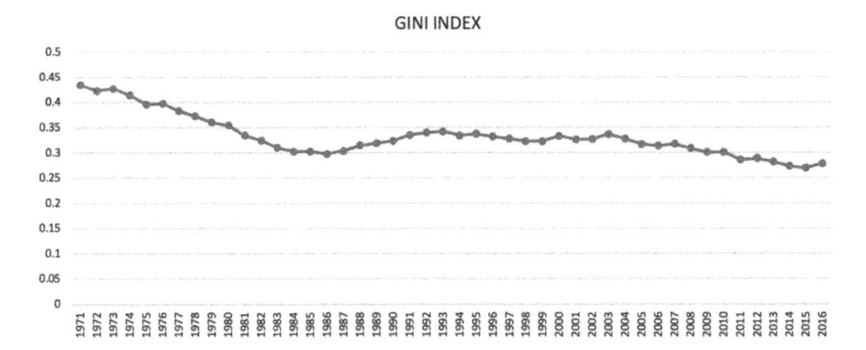


*Figure 5.3* *Gini index on $CO_2$ emissions, 1971–2016*

*Source:* Our estimation based on IEA, Indicators for $CO_2$ Emissions, data extracted on 21 December 2018.

Estimation of the Gini index confirms that the reduction in the gap also concerns per capita $CO_2$ emissions, other than per capita energy consumption as seen above. We can conclude that the growth in total $CO_2$ emissions in North Africa and Middle Eastern countries is due not only to population growth, and increase in per capita energy consumption, but especially to the increasing share of fossil fuel consumption. The growth in North Africa and the Middle East, and contrastingly, the reduction in European countries, is leading to a reduction in the divide between developed and emerging countries within the Mediterranean region.

## 5. ENERGY DEPENDENCY AND ENERGY SECURITY

The availability of large quantities of energy represents an indispensable condition for both advanced and emerging economies. For this reason, there is stiff competition among different countries to secure their energy supplies, world energy consumption having more than doubled in the last fifty years. The increase has been driven especially by China, which became the major world's oil importer in 2013–2014. In the Mediterranean region, the problem of energy security is particularly severe because a large increase in energy demand is expected both for Europe, North African and Middle Eastern countries. In addition, the reduction in production and exports from Libya, the war in Syria, and political and social turmoil in Algeria, as well as in Egypt and Lebanon, raise the problem of energy dependency to the top of the political agenda.

Energy security means secure, stable and affordable supplies, and it cannot be pursued without political, economic and social implications (Clô, 2016). Terrorist attacks, political instability of oil-producing countries, interruption of oil and gas trade flows, and high volatility of energy prices, have all contributed to energy "insecurity" in the Mediterranean area. As shown by Jalilvand and Westphal (2018), in securing energy supplies, interactions between exporters, importers, producers and consumers are of key importance, as are regional equilibria among producers. For the Mediterranean area, Gaddafi's collapse was a catastrophe for Algeria (Oumansour, 2019). When NATO began bombing Libya in 2011, on the instigation of France, Algeria was opposed to such a course of action, fearing that the end of the Gaddafi regime would lead to uncontrollable regional instability. Although Gaddafi competed with Algeria for influence in sub-Saharan Africa and the Sahara itself, the Algerian authorities supported him until the end, as his regime guaranteed the stability of the Maghreb and the Saharan belt, as well as being an expression of the same ideological and political strain. In 2019, Algeria was shaken by a national protest movement calling for radical political change, triggered by the announcement by President Abdelaziz Bouteflika, in office since 1999, of wanting to run for his fifth consecutive presidential term. Bouteflika resigned on 2 March after six weeks of peaceful demonstrations. The postponement of the vote gives us a measure of how deep the political crisis is.

The collapse of oil prices since 2014 and the increase in the price of basic necessities have particularly affected the working classes and have accentuated social inequalities. In addition, a new social class has emerged, formed by educated young people, with the right to vote, who aspire to change, especially after news on the political corruption of the country spread thanks to the internet. The Algerian power system is very complex and cannot be fully understood unless one considers the country's colonial past, marked by French domination which ended only after a long and bloody war, from 1954 to 1962, resulting in hundreds of thousands of victims. Indeed, "the history of Algeria is a history of France and vice versa" (Oumansour, 2019), since the French colonization of Algeria lasted more than a century.

Algeria is experiencing a delicate phase not only in political terms, but also from the economic perspective. The Algerian economy is dominated by the State and it is essentially based on hydrocarbon exports, which in 2016 represented about one-third of total government revenues and about 90% of total exports. The high dependence on crude oil exports has increased the vulnerability of Algeria's economy to crude oil price volatility. Following the Organisation of the Petroleum Exporting Countries (OPEC) production cut agreement in 2016, Algeria had been able to comply with its obligation to limit its production to 1.04 million b/d.[26]

As for natural gas exports, only about 15% was exported to the Middle East and North Africa, while almost all of it, about 83% in 2016, was sent to Europe, with Italy and Spain being the main importers.

As for Libya, in 1969, a coup allowed the young colonel Gaddafi to come to power. One of Gaddafi's first moves was to start the nationalization of hydrocarbons, the oil and gas sector, until then mainly in the hands of international companies, especially the USA. The backlash was immediate, as production was reduced from three million barrels a day to one million barrels a day in the 1980s, to then stabilize around an average of 1.5 million (Bellodi, 2019). Based on the above, we can assume that the oil shock of 1973–74 was caused not only by the oil embargo decreed by the Arab countries, but also by the profound changes that were taking place in Libya.

Libyan oil was strategic for Europe for several reasons. First of all, it is of good quality, being light and very low in sulphur, as well as inexpensive. Furthermore, importing oil from Libya did not present the risks that oil imports from the Middle East entailed, being transported through the Suez Canal or through expensive oil pipelines, which were often very risky due to continuous geopolitical tensions.

The interruption of oil imports from Libya represented a great shock for energy markets, and Europe especially had to diversify its supply sources. These aspects reflected on the dynamics of oil prices. Italy, which imported mainly from Libya, had to replace Libyan oil with oil from Azerbaijan.

Europe has always had to grapple with the problem of energy dependency, which has conditioned its political relations with energy-producing and -exporting countries since World War II. As was seen above, energy was an important factor in the Cold War, but the problem of Europe's energy dependency on the Middle East and Russia has even affected its alliance with the USA. At present, Europe continues to be highly dependent on energy imports from Russia. In 2017, EU-28 had an energy dependency of 55%, which means that more than half of the energy consumed in the EU is imported from non-EU countries. Russia is the main EU supplier of crude oil, natural gas and coal. Indeed, in terms of the geography of energy supply, 30.03% of crude oil, 39.8% of natural gas and 38.8% of solid fuels were imported from Russia in 2017.[27]

As for oil, the second partner in order of importance is Norway, with a share of 11.4% (see Figure 5.4). About 15% was imported by Middle East countries, namely Iraq (8.2%) and Saudi Arabia (6.6%). Dependence on North Africa strongly declined over time due to the Libyan crisis, and recent fighting in Libya has caused another interruption of oil exports.

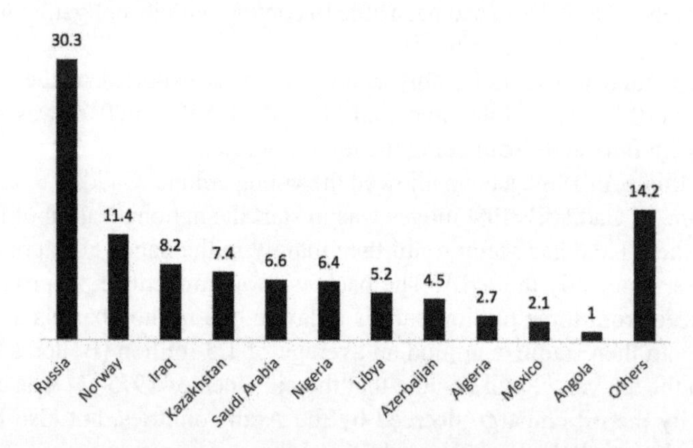


*Figure 5.4 EU imports of crude oil by partners in 2017 (%)*

*Source:* Eurostat, data extracted on 8 October 2019 from http://ec.europa.eu/eurostat.

As for natural gas, the main partner is Russia, but the role of Algeria also continues to be important: it supplied 10.7% of total EU imports of natural gas during 2017 (Figure 5.5).

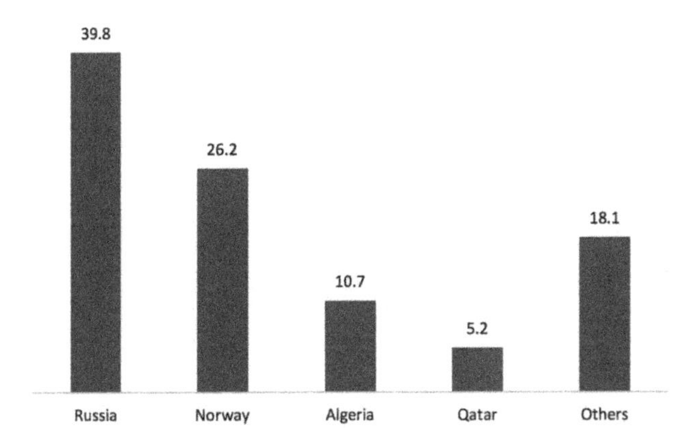


*Figure 5.5* *EU imports of natural gas by partners in 2017 (%)*

*Source:* Eurostat, data extracted on 8 October 2019 from http://ec.europa.eu/eurostat.

Within the Mediterranean region, diversity prevails also in terms of energy dependency (see Figure 5.6). It is hardly surprising that those countries that produce nuclear electricity have lower levels of energy dependency, under the EU average, namely France (48.6% in 2017) and Slovenia (50.4%), while countries that do not produce nuclear energy have higher levels of energy dependency. The highest levels are reached by Malta, with an energy dependency of 102.9%, and Cyprus, with 96.3%. Malta is completely dependent on oil, which represents about 96% of its total energy consumption, while only the remaining 4% is represented by renewable energy. Also Cyprus is completely dependent on oil, which in 2016 represented 93% of its total primary energy consumption, while renewable energy accounted for only 7%.

In the other European countries of the Mediterranean region, the levels of energy dependency exceed 70%. With respect to Italy, the levels are slightly lower in Spain because the latter produces nuclear electricity.

Over time, there has been a change in the geography of energy supply, resulting from the emergence of new exporting countries, the decline of former exporting countries, and in general from the dynamics of world energy demand and supply (Bartoletto, 2016b; 2012b).

In 1990, about 54% of Italian oil imports came from Africa. The main supplier was Libya, which accounted for about 24.5 million tons (Mt), followed at some considerable distance by Egypt (6.2 Mt) and Algeria (4.6 Mt). At present, the role of Africa as Italy's oil supplier has greatly diminished. Indeed, in 2017, only 18% of imports came from Africa, and only 5 Mt were

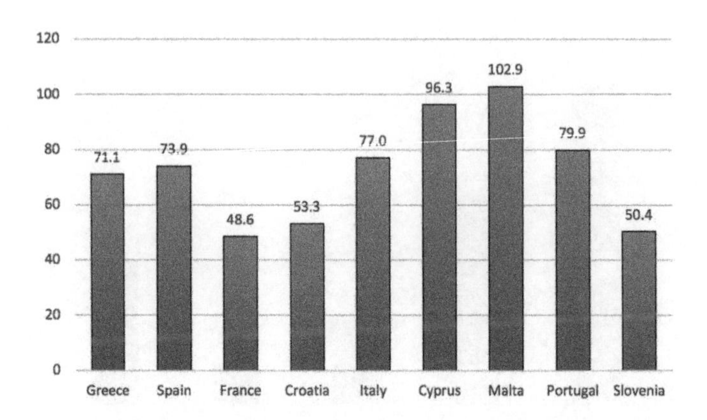


*Figure 5.6* *Energy dependency in European countries in 2017 (%)*

*Source:* Eurostat, Energy imports dependency, data extracted on 8 October 2019 from http://ec.europa.eu/eurostat.

supplied by Libya (Unione Petrolifera, 2018). On the other hand, imports from the Middle East have increased, with Iran, Iraq and Saudi Arabia the main oil suppliers. In the past the role of the Middle East as an oil supplier had declined to the advantage of Libya. However, the crisis in the latter country has allowed the Middle East to recover its position, as emerges from analysis of the data in question. Indeed, Italian oil imports from the Middle East, after the peak of 34.5% in 2005, decreased to a minimum of 23.8% in 2014. From then on, they rapidly increased, reaching a peak of 41.6% in 2017. In addition, the growth of oil imports from the former Soviet Union has been very impressive: its share increased from about 8% in 1990 to 34% in 2017. Within this trading area, the main partner is Azerbaijan, which in 2017 provided 12.4 million tons of oil, followed by Russia, with 6.5 million tons (Unione Petrolifera 2018).[28] At present, Azerbaijan is Italy's main oil supplier.

The issue of security of energy supplies has been threatened by the high volatility of oil prices. During 2018, the average price of oil increased by 30% due to the alliance between OPEC and non-OPEC countries. The high volatility of oil prices is influenced by contrasting factors. On the one hand, there has been a reduction in demand due to the slowdown in global growth and the increased energy independence of the USA; at the same time, there has been a reduction in supply due to conflict and geopolitical tension. During 2018, about 56% of growth in world oil demand was due to increased demand from China and India. On the supply side, oil production has grown, especially due to the rise of US production of *tight oil*.

Within the Mediterranean region, Turkey may be seen as a separate case due to its role as an energy corridor and its proximity to Europe, Central Asia and Middle Eastern countries such as Iran, Iraq and Syria. Turkey is not only an important energy transit, but is also an emerging economy with rapid growth in energy demand, and thus security of energy supply occupies a key position in its political strategies. Turkey is a non-Arab country in the Middle East (Scholl, 2018), but Turkey also shares borders with Europe and Central Asia, and is on the list of candidate countries for EU membership because it plays a strategic role for the EU in terms of energy supplies and other issues related to security and migration. However, in June 2018 the General Affairs Council decided that accession negotiations would be frozen because Turkey had failed to respect the fundamental rights of peoples and principles of democracy, while Turkey's invasion of Syria has further contributed to stalling its inclusion among EU member states. Thus also the role of Turkey as an energy corridor has been impacted over time not only by energy trends, but also by changing dynamics in the Middle East and neighbouring areas. Despite its proximity to oil-rich countries in the Middle East, Turkey is not well endowed with energy sources, and its level of energy dependency has grown over time. Like North African and Middle East countries, Turkey has experienced a rapid growth in energy consumption and, at world level, it is one of the countries with the highest growth rates. Fossil fuels dominate the energy balance, while renewable energy sources play a very limited role, with the exception of hydroelectricity. This means that about 90% of energy demand is satisfied through imports, with severe negative effects on the balance of payments. Thus Turkey is greatly exposed to energy price volatility, while the reduction in oil prices since 2014 has only partially benefited its national accounts. Another important aspect to consider is that energy intensity continues to rise, which means that the growth of gross domestic product (GDP) and energy consumption are not yet decoupled, as they are in more advanced economies.

To summarize, growing energy consumption, energy intensity and $CO_2$ emissions are the main features of Turkish energy policies that do not exclude an expansion of the role of coal, being less costly than oil and gas. Turkey has large coal reserves, but much consists of low-quality lignite which is unsuitable for certain industrial uses. Most hard coal is therefore imported from Colombia, Russia, South Africa and Australia. Another strategy in Turkey's energy policy is to expand the role of gas, whose consumption has more than doubled in the last twenty years. The problem is that growing demand is reaching the limits of "transmission infrastructure and import capacity" (Scholl, 2018). Gas consumption increased rapidly after Turkey signed its first gas sale contract with Russia in 1986. At present, Russia continues to represent the main gas supplier for Turkey, and according to an IEA report,[29] about 55% of imports come from Russia through the Trans-Balkan Pipeline and Blue

Stream Pipeline. Other gas suppliers are Iran, with a share of 16%, Azerbaijan, which supplies about 12%, and Algeria, from which Turkey imports about 8%. The Turkish strategy to expand the role of gas in the energy balance is strictly linked to the EU's strategy, as the "Southern Gas Corridor" (SGC), which represents an important component of the cooperation of Turkey and Europe in the energy sector. The project should involve the construction of several gas pipelines that transport gas from the Shah Deniz field to Turkey and then to Europe. Shah Deniz is Azerbaijan's largest gas field, located in the South Caspian Sea. Another important project linking Turkey and Europe is the Trans-Anatolian Pipeline (TANAP), which is a gas pipeline that reaches Europe from Azerbaijan, through Georgia and Turkey. Moreover, Turkey shares with Europe only part of its energy strategies, because another part is shared with Russia. While Europe is trying to reduce its energy dependency on Russia, Turkey is of key importance for Russia with a view to developing a new network of gas pipelines that may allow a reduction in gas exports to Europe through the Ukraine. Thus the picture of Turkish energy strategies is complex, and the discovery of new gas fields in the Eastern Mediterranean, such as Zohr off Egypt's shores, and the fields of Aphrodite, Tamar and Leviathan off the coasts of Israel and Cyprus, contribute to make energy dynamics ever more complex since Turkey could be an export market and a transit country for such gas flows. Moreover, Turkey's position with respect to the war in Syria and Libya, together with growing internal political instability, risks compromising planned energy projects. Thus making predictions on Turkey's energy policies and its role in energy markets is far from straightforward.

Fighting in Libya and Syria has seen the great powers such as the USA, Russia and Europe aligned with divergent positions and interests. Yet judging from recent events it seems that war and peace in these two countries will be decided by Erdoğan's Turkey and Putin's Russia. The parliament in Ankara has approved the dispatch of soldiers to Tripoli against General Khalifa Haftar, with the aim, as announced by Erdoğan, to guarantee the survival of the legitimate government of Fayez al-Sarraj. However, the Egyptian government strongly condemned the vote of the Turkish Parliament authorizing the sending of troops to Libya in support of the government of Fayez al-Serraj. The Cairo foreign ministry stressed that the deployment of Turkish troops could "have a negative impact on the stability of the Mediterranean region" and asked the international community to react immediately. Furthermore, according to Egypt, the Turkish Parliament's decision is a violation of the UN Security Council resolutions on Libya. Contrasting interests divide Egypt and Turkey, and energy is one of these. Recently Erdoğan signed the "Mediterranean pact" with Libya, with which he claims the exploitation of offshore gas resources in the exclusive area of Greek Cyprus in competition with ENI and Total. Putin and Erdoğan now decide together the fate of Syria, Libya and also, in part, those of gas supplies, having just inaugurated the Russian–Turkish pipeline of the Turkish Stream, in explicit opposition to the East-Med gas pipeline projects between Egypt, Israel, Cyprus and Greece (Negri, 2019). Italy is observing the situation with concern as Tripolitania contains ENI's major energy interests, such as the Greenstream gas pipeline. In the meantime, statements from USA, Iran and Iraq fuelled tensions in the Middle East after the US drone strike that killed an important Iranian military commander, General Soleimani, on 3 January 2020. On 5 January 2020, the US President Donald Trump threatened to impose sanctions on Iraq, the second largest OPEC producer, in the event of a forced withdrawal of US troops from the country. Baghdad has called for the USA and other military forces to leave Iraq.[30] Trump also declared that the USA will target other Iranian sites if Tehran attacks US citizens or assets in response to the killing of General Soleimani. Consequently, the problem of security of energy supplies, instead of being resolved, has worsened in recent months due to the strong tensions existing within and between the Middle East, USA and Russia, impacting strongly on the Mediterranean region, which needs to find new energy strategies and a new equilibrium within a very uncertain and constantly changing geopolitical picture.

## NOTES

1. Nordhaus (1992, p. 28).
2. Nordhaus (1992, p. 39).
3. Barcelona declaration adopted at the Euro–Mediterranean Conference, 27–28 November 1995.
4. Council Regulation (EC) No. 2698/2000 of 27 November 2000 amending EC Regulation 1488/96.
5. Council Decision of 29 May 2006 on the conclusion by the European Community of the Energy Community Treaty (2006/500/EC).
6. For further details on the Energy Community Treaty and participating countries, see Quadri (2012).
7. For more information see http://www.desertec.org/en/organization/.
8. https://ec.europa.eu/energy/en/topics/international-cooperation/eu-cooperation-other-countries/neighborhood-south, last accessed 7 April 2019.
9. EU, Dialogue énergétique de haut niveau entre l'Algérie et l'Union Européenne, Algiers, 5 May 2015.
10. Memorandum of understanding on a strategic partnership on energy between the European Union and the Arab Republic of Egypt 2018–2022 (last accessed 7 April 2019).
11. European Parliament (2009), Directive 2009/28/EC of the European Parliament and of the Council of 23 April 2009 on the promotion of the use of energy from renewable sources and amending and subsequently repealing Directives 2001/77/EC and 2003/30/EC.
12. European Commission (n.d.), Renewable energy directive, https://ec.europa.eu/energy/en/topics/renewable-energy/renewable-energy-directive/overview, last accessed 25 February 2020; European Union (2018), Directive 2018/2001 of the

European Parliament and of the Council of 11 December 2018 on the promotion of the use of energy from renewable sources.
13. European Union (n.d.), Use of international credits, https://ec.europa.eu/clima/policies/ets/credits_en, last accessed 24 February 2020.
14. Emission trading system, Market stability reserve in https://ec.europa.eu/clima/policies/ets/reform_en, last accessed 7 April 2019.
15. European Union (n.d.), Use of international credits, https://ec.europa.eu/clima/policies/ets/credits_en, last accessed 24 February 2020; European Commission (2019), The European Green Deal, Brussels, 11 December 2019.
16. https://sustainabledevelopment.un.org/partnership/.
17. IEA, *World Energy Balances*, data extracted on 19 April 2019 (reference year 2016); US Energy Information Administration, Country analysis brief: Turkey, February 2017, at www.eia.doe.gov.
18. US Energy Information Administration, Country analysis brief: Turkey, October 2006 in www.eia.doe.gov.
19. US Energy Information Administration, Country analysis brief: Egypt, 2 June 2015 in www. eia.doe.gov.
20. For further details, see https://www.medener.org/.
21. https://ec.europa.eu/clima/policies/strategies/2030_en.
22. IEA, Indicators for $CO_2$ Emissions, data extracted on 21 December 2018.
23. Our calculations from IEA, Indicators for $CO_2$ Emissions, data extracted on 21 December 2018.
24. In December 2018, the revised renewable energy directive 2018/2001/EU entered into force (European Commission (2018), Renewable Energy – Recast to 2030 (RED II), https://ec.europa.eu/jrc/en/jec/renewable-energy-recast-2030-red-ii (last accessed on 25 February 2020).
25. IEA, *World Energy Balances*, data extracted on 19 April 2019; US Energy Information Administration, Country analysis briefs: Turkey, February 2017, at www.eia.doe.gov.
26. US Energy Information Administration, Country Analysis Executive Summary: Algeria, 25 March 2019.
27. European Union (2019), *Shedding light on energy in EU. A guided tour of energy statistics*, May 2019.
28. For Italian oil imports from Africa, Middle East and former Soviet Union, see Unione Petrolifera (2018).
29. US Energy Information Administration, Country analysis brief: Turkey, February 2017, at www.eia.doe.gov.
30. Hall (2020), "No, no America": thousands march in Iraq to demand exit of US forces, Friday 24 January 2020.

# Conclusions

The Mediterranean region has always played a strategic role in energy supplies, a role that has become increasingly important since World War II. Post-war oil crises, such as the Suez crisis, the 1970s oil crises, up to the Gulf crisis, have shown the central role of the Mediterranean region for security of energy supplies. Also in the Cold War and in the Marshall Plan energy security has been a major issue. The ongoing war in Syria and especially the conflict in Libya have shown once again the strategic importance of the Mediterranean region in terms of energy geopolitics.

World energy consumption more than doubled between 1971 and 2016. In 2017, global energy demand increased significantly compared to 2016, mainly driven by growth demand in non-OECD countries, even though, during 2018, there was acceleration in energy demand in OECD countries. According to the scenarios for the world economy to 2060 the weight of emerging economies will rise and living standards will continue to improve.[1] This means that, without a new energy transition toward a low-carbon energy system, fossil fuel consumption and $CO_2$ emissions will continue to rise, with severe consequences for climate change. In addition, growing energy consumption is causing more competition between different countries to secure energy supplies. Also within the Mediterranean region, energy consumption has more than doubled over the past 50 years, due to the increase in population and economic development. The Mediterranean is an area characterized by strong geographical, institutional and economic-social differences. In recent decades there has been a process of economic convergence, which has also been reflected in the trend of per capita energy consumption and $CO_2$ emissions, as shown by our estimates of the Gini index. Despite the convergence process, there remain great differences in income, energy consumption and $CO_2$ emissions levels: in North Africa, energy consumption has increased about tenfold, while in European countries there has been a reduction in growth rates, also due to the financial crisis of 2007–2008. Therefore, over the past 50 years there has been a major change in the distribution of energy consumption within the Mediterranean region: while in the early 1970s North Africa represented only 4% of the total consumption of the Mediterranean region, and the countries belonging to the European Union (EU) 81%, during 2016 the relative weight of North Africa rose to 19% while that of the EU decreased to 59%. There has been an increase not only in overall consumption but also in per capita consumption in the countries of North Africa and the Middle East, although the gap compared to the countries of the EU remains very wide.

In Turkey, strong economic growth since the mid-1990s has had important repercussions on energy consumption and the environment, leading to a significant increase in the consumption of fossil fuels and $CO_2$ emissions. Even in 2016, Turkey was the country with the highest levels of $CO_2$ emissions within the Mediterranean region.[2] Analysis of the past 50 years has shown that growth rates of $CO_2$ emissions in North Africa are very high, especially in Algeria and Egypt. The same can be said for the Middle Eastern countries, especially Jordan and Lebanon, where both the increase in per capita consumption and the increase in the population are driving the growth of $CO_2$ emissions.

The problem of climate change ranks uppermost on the political agenda of many governments, and from this perspective the Mediterranean requires special attention, as the strong growth in energy demand on the part of North African and Middle Eastern countries will impact further on the environment. In the Mediterranean area, renewable energies represent around 11% of total primary energy consumption, while fossil fuels represent 75%, and nuclear power 12%. The main source of energy remains oil, and although its consumption has halved over the years, it still accounts for 37% of primary energy consumption. The share of natural gas grew from 6% in 1971 to 28% in 2016, while that of coal decreased from 17% to 12% over the same period. As for nuclear power, its production within the Mediterranean region is concentrated in three countries: 86.23% of the total is produced in France, 12.54% in Spain, and 1.22% in Slovenia in 2016.

Over the last decade, there has been significant progress in the renewable energy sector, but significant differences remain between the different areas, and within these, between the individual countries that make up the Mediterranean region. Strong acceleration in the production and consumption of renewable energy occurred after the signing of the Kyoto Protocol which fuelled an expansionary phase of investments in renewable energies. The latter became profitable thanks to public subsidies and the priority of provision guaranteed to renewables. An important role has also been played by technological progress, which allows a reduction in generation costs of renewables. However, also with regard to renewables, there are profound differences between the countries belonging to the EU and those belonging to other areas. Growth has been particularly intense in EU countries, while in North Africa and the Middle East, where oil and natural gas consumption are increasing rapidly, renewable energy continues to play a very marginal role.

However, there are major differences between the various countries belonging to the same area. In North Africa, Tunisia represents the country with the highest share of renewable energy, equivalent to 11% of total primary energy consumption in 2016. The development of renewable energy in Tunisia is based on the "Tunisian Solar Plan", which started in 2009 and was reviewed in 2015. The plan sets a target for the share of renewables as 30% of total electricity production.

Morocco is also intensifying its efforts in favour of renewables, even if its policies present major contradictions, since at the same time it continues to import increasing quantities of hard coal. From 2000 to today, Morocco's gross domestic product (GDP) has more than doubled and in 2017 the economy grew by about 4%. The country is carrying out important reforms aimed at developing industry, infrastructure, agriculture and the energy sector, thanks to the attraction of foreign capital. The choice of renewable energies is a mandatory choice, as Morocco has no reserves of oil and natural gas, and therefore must cover about 90% of its energy needs with imports. Increasing the share of renewables is an obligatory step to reduce the high energy dependence on fossil fuels. In contrast, Algeria satisfies its entire demand for energy by consuming oil and natural gas, thanks to the large reserves available which allow it to be also the foremost exporter in the Mediterranean area. In Egypt, renewables make up only 4% of energy demand, and although the country has announced that it wants to develop renewables, the greatest efforts are concentrated in the production of fossil fuels, especially natural gas, efforts that in the recent years have led to the discovery of new gas fields. The situation in the Middle East countries is not very different. In Jordan, renewables represent just 3%, and in Israel and Lebanon they represent only 2%.

Despite the increase in the share of renewables in countries belonging to the EU, within the Mediterranean region as a whole $CO_2$ emissions have doubled, growing from 1049 Mt in 1971 to around 2017 Mt in 2016. Still considering absolute values, almost half of such emissions are produced by the countries belonging to the EU, especially by France, Italy and Spain, which together produce about 42% of the total emissions of the Mediterranean area. This fact should not be overlooked: while the growth rates of $CO_2$ emissions in the countries belonging to the EU have decreased significantly over time, the fact remains that they are responsible for most of the emissions of the entire Mediterranean region. In the meantime, since $CO_2$ emissions are growing in North Africa and the Middle East, the estimate of the Gini index confirms that the gaps are narrowing, not only for per capita energy consumption, but also for $CO_2$ emissions. This aspect should not be underestimated in the political agenda of governments, and especially in Europe. The Mediterranean area is particularly exposed to climate change. Just think of the Sahara Desert, whose area has grown considerably over time. Climate change in sub-Saharan Africa, especially changes in rainfall and temperatures, have had a negative impact on crops, helping to fuel hunger, cross-border migration and intra-regional conflict. Climate change is also aggravating the problem of water shortages, which already afflict many African cities with high population density. The risk is that in the coming years new conflicts will be triggered by the lack of food and water. If there is no remedy, the impact on North Africa will be disastrous, with serious consequences also for the countries of Southern Europe, especially in terms of migratory flows. In North African countries, the population is growing rapidly, together with the number and size of cities, and this will lead to a sharp increase in energy consumption and $CO_2$ emissions in the coming decades. For this reason, a greater effort is needed to increase the share of renewable energy, especially in such countries where the increase in energy consumption will be greatest, not only due to the increase in population, but also for the development of the industrial sector and other sectors of the economy. The EU has tried in different ways and at different times to initiate dialogue and cooperation with the countries of North Africa and the Middle East belonging to the Mediterranean. A fundamental step was represented by the Euro–Mediterranean Conference in Barcelona in November 1995, making it possible to define the political, economic and social framework within which to develop such relations. After a long period of stagnation, in 2008 Euro–Mediterranean cooperation was relaunched, with the creation of the Union for the Mediterranean (UfM), created on 13 July 2008 during the French presidency of the EU. The UfM is an intergovernmental institution in which the 28 member countries of the EU and 15 countries on the southern and eastern shores of the Mediterranean participate with the aim of promoting dialogue and cooperation. All countries in the Mediterranean region are part of the UfM, except for Syria and Libya. However, the riots that broke out after the Arab Spring of 2011 highlighted the fragility of such organizations compared to the political, social and economic complexity of the countries in question. The recent epilogue of the war in Syria and Libya which, at the time of writing, sees Russia and Turkey prevail as the main players in these two countries, has shown that European politics is still too weak compared to other areas of the Mediterranean.

In addition, the reduction in oil prices since 2014 is worsening the fiscal imbalances and public debt of countries that rely mainly on revenues from oil exports, with serious consequences also for internal stability and economic and international relations. This is an important aspect that should be taken into consideration when policymakers think about the development of Euro–Mediterranean relationships. Economic history has shown that for a country well endowed with energy resources, long-run dependence on the exploitation of these natural resources is not sustainable. As stressed by Fouquet (2019), economic growth based exclusively on the exploitation of natural resources should be only a temporary phase in the economic history of a country, because otherwise it risks limiting the country's development potential. On the contrary, the rents deriving from exports of energy sources should be used to diversify the economy, financing the development of other economic activities that do not depend on the exploitation of natural resources. Europe has a key role to play, giving new strength to the Euro–Mediterranean partnership, which could be an excellent tool for restoring peace and stability to the region. We have analysed the problem of energy dependency and we have seen that Europe depends on the Mediterranean region for its energy needs. However, Europe should implement Euro–Mediterranean relations not only to secure energy supplies, but also to promote the economic development of these countries, helping them to develop industries and economic activities that are not based on the exploitation of natural resources. In the long term, a growth not based exclusively on natural capital exploitation is fundamental both in terms of economic development and for the development of democratic principles. In oil- and gas-producing countries, dictatorial regimes and power groups are known to derive their strength from hydrocarbon export income, and from this perspective, the economic history of Libya represents an emblematic example. The implementation of policies to reduce $CO_2$ emissions continues to clash with the serious tensions that characterize the Mediterranean region. The presence of oil and natural gas reserves, and of important oil and gas pipelines connecting exporting and importing countries, represent fundamental issues that see international powers ranged against each other for the different interests at stake. The war in Libya and Syria, and the as yet unresolved issue between Israel and Palestine, make the Mediterranean region highly unstable. Although the EU is seeking to intensify relations with North Africa and the countries of the Middle East in the energy sector, concrete implementation of energy policies faces many obstacles. The gap between what is announced and what is actually achieved remains wide. Since the Mediterranean is one of the regions most exposed to the problem of climate change, the increase in the share of renewables is above all a question of greater environmental sustainability of energy consumption, but it is also an important aspect of the security of energy supplies. Unfortunately, following the global economic and financial crisis of 2008, the recovery of economic growth has become the main priority on the political agenda of various governments, and the environmental issue has dropped down the agenda, as seen from the reduction of investment rates in renewable energy in some European countries and the USA.[3] The economic crisis, and in other cases the lower cost of fossil fuels, have slowed down or cancelled the development programmes for renewables. Unfortunately, economic growth and environmental policies are still not viewed as complementary as they should be, and the world economy is still far from the "Green Growth" paradigm, which gives economies the possibility to develop and grow without damaging the environment, creating new jobs through "green investments" (Fouquet, 2019; Fouquet and Hippe, 2019). Once again, economic history can point us in the right direction, especially thanks to the study of past energy transitions which have played a decisive role in industrial revolutions. In analysing the productivity of energy over the nineteenth and twentieth centuries, we saw that fossil fuels (modern sources) in the initial phase were less efficient than traditional sources such as wood. The progressive replacement of renewable with non-renewable energy resources occurred at different rates in different countries. The transition from wood to coal was a slow process since the two sources coexisted for a long time. Up until the early decades of the twentieth century, much of Western Europe still depended significantly on traditional sources of energy. While the energy transition occurred very quickly in Northern Europe, especially in England, unlike in southern Europe, traditional sources of energy still represented 70–80% of the energy consumed in the early twentieth century. In Italy and Spain in particular, the contribution of traditional energy sources decreased to less than half of the total primary consumption only on the eve of World War II (Bartoletto and Rubio-Varas, 2008).

Economic history teaches that each energy transition has a cost, generally associated with lower energy efficiency of the new energy sources. However, over time, mainly thanks to technological progress, this constraint can be overcome. In addition, it should not be forgotten that the depletion of natural capital and climate change have a cost, as shown by several studies that attempt not only to quantify the external costs of air pollution (Fouquet, 2011; Matus et al., 2012), but also to show, through empirical methods, that climate change has been and will continue to be a constraint for economic development, other than a challenge for health and survival.

Policymakers need to learn the lesson of economic history on past energy transitions to better manage a new energy transition towards low-carbon growth, which means a new industrial revolution, in which a low-carbon energy system is an integral part (Fouquet, 2017). The protection of natural capital is fundamental for economic growth, something which policymakers should always bear in mind.

## NOTES

1. OECD (2018); IEA (2019), *World Energy Balances*.
2. IEA (2018b), Indicators for $CO_2$ Emissions, data extracted on 21 December 2018.
3. Bloomberg New Energy Finance (2018).

# Statistical appendix

*Table 1A.1* *Per capita GDP in Mediterranean countries, 1971–2016 (2010 US dollars)*

| Time/Country | 1971 | 1980 | 1990 | 2000 | 2008 | 2011 | 2016 |
|---|---|---|---|---|---|---|---|
| Albania | 3500 | 4741 | 4576 | 5484 | 9034 | 9931 | 10931 |
| Algeria | 6540 | 10249 | 10035 | 10000 | 12390 | 12734 | 13640 |
| Bosnia-Herz. | – | – | 1533 | 6000 | 9316 | 9459 | 10914 |
| Croatia | – | – | 14250 | 15114 | 21159 | 19651 | 20405 |
| Cyprus | 5833 | 14400 | 22000 | 29286 | 34625 | 34625 | 32625 |
| Egypt | 2969 | 4426 | 5791 | 7239 | 9112 | 9625 | 10110 |
| Macedonia | – | – | 9550 | 8700 | 10810 | 11381 | 12857 |
| France | 18603 | 23926 | 29007 | 34103 | 36815 | 36623 | 37184 |
| Greece | 14955 | 19938 | 20107 | 24398 | 31234 | 25676 | 23722 |
| Israel | 13767 | 15923 | 19106 | 25540 | 27757 | 29654 | 32000 |
| Italy | 17495 | 23936 | 30183 | 35427 | 36542 | 34795 | 33561 |
| Jordan | 4889 | 7500 | 6083 | 7098 | 9508 | 9000 | 8179 |
| Lebanon | 13208 | 10154 | 7704 | 12406 | 14341 | 15348 | 13017 |
| Libya | 62136 | 54406 | 25545 | 21296 | 28180 | 10952 | 7143 |
| Malta | 5000 | 12667 | 14000 | 23500 | 28500 | 29250 | 38250 |
| Montenegro | – | – | – | – | 14500 | 14500 | 15833 |
| Morocco | 2368 | 3030 | 3863 | 4433 | 6073 | 6641 | 7244 |
| Portugal | 11322 | 14818 | 20200 | 26068 | 27575 | 26764 | 27291 |
| Serbia | – | – | 5426 | 7062 | 12216 | 12417 | 13000 |
| Slovenia | – | – | 18300 | 21900 | 30500 | 27286 | 28524 |
| Spain | 14413 | 17905 | 23120 | 29463 | 33583 | 31582 | 32774 |
| Syria | 2697 | 4697 | 4306 | 5201 | 5951 | 5273 | 1842 |
| Tunisia | 3327 | 4984 | 5524 | 7423 | 9808 | 9880 | 10535 |
| Turkey | 7014 | 8070 | 10806 | 13253 | 17180 | 18961 | 23483 |
| Yugoslavia | 6300 | 9546 | – | – | – | – | – |

*Source:* Our calculations based on IEA, Indicators for $CO_2$ Emissions, data extracted on 21 December 2018.

*Table 1A.2 Population in Mediterranean countries, 1971–2016 (millions)*

| | 1971 | 1980 | 1990 | 2000 | 2008 | 2011 | 2016 |
|---|---|---|---|---|---|---|---|
| | | | **European Union** | | | | |
| Cyprus | 0.6 | 0.5 | 0.6 | 0.7 | 0.8 | 0.8 | 0.8 |
| Croatia | – | – | 4.8 | 4.4 | 4.4 | 4.3 | 4.2 |
| France | 52.4 | 55.2 | 58.2 | 60.9 | 64.3 | 65.3 | 66.9 |
| Greece | 8.9 | 9.7 | 10.3 | 10.8 | 11.1 | 11.1 | 10.8 |
| Italy | 54.1 | 56.4 | 56.7 | 56.9 | 59.2 | 60.1 | 60.6 |
| Malta | 0.3 | 0.3 | 0.4 | 0.4 | 0.4 | 0.4 | 0.4 |
| Portugal | 8.7 | 9.9 | 10.0 | 10.3 | 10.6 | 10.6 | 10.3 |
| Slovenia | – | – | 2.0 | 2.0 | 2.0 | 2.1 | 2.1 |
| Spain | 34.6 | 38.0 | 39.3 | 40.6 | 46.0 | 46.7 | 46.5 |
| **Total** | **159.6** | **170.0** | **182.3** | **187.0** | **198.8** | **201.4** | **202.6** |
| | | | **Candidates EU** | | | | |
| Albania | 2.2 | 2.7 | 3.3 | 3.1 | 2.9 | 2.9 | 2.9 |
| Bosnia-Herz. | – | – | 4.5 | 3.8 | 3.8 | 3.7 | 3.5 |
| Montenegro | – | – | – | – | 0.6 | 0.6 | 0.6 |
| Macedonia | – | – | 2.0 | 2.0 | 2.1 | 2.1 | 2.1 |
| Serbia | – | – | 10.1 | 8.1 | 7.4 | 7.2 | 7.1 |
| Turkey | 36.2 | 44.4 | 55.1 | 64.3 | 71.1 | 74.0 | 78.2 |
| **Total** | **38.4** | **47.1** | **75.0** | **81.3** | **87.9** | **90.5** | **94.4** |
| | | | **Middle East** | | | | |
| Jordan | 1.8 | 2.4 | 3.6 | 5.1 | 6.5 | 7.6 | 9.5 |
| Israel | 3.0 | 3.9 | 4.7 | 6.3 | 7.4 | 7.8 | 8.5 |
| Lebanon | 2.4 | 2.6 | 2.7 | 3.2 | 4.1 | 4.6 | 6.0 |
| Syria | 6.6 | 8.9 | 12.4 | 16.4 | 20.3 | 20.9 | 18.4 |
| **Total** | **13.8** | **17.8** | **23.4** | **31.0** | **38.3** | **40.9** | **42.4** |
| | | | **North Africa** | | | | |
| Algeria | 15.0 | 19.3 | 25.9 | 31.2 | 34.9 | 36.8 | 40.6 |
| Egypt | 35.9 | 44.1 | 57.4 | 69.9 | 81.0 | 85.9 | 95.7 |
| Libya | 2.2 | 3.2 | 4.4 | 5.4 | 6.1 | 6.2 | 6.3 |
| Morocco | 16.3 | 20.0 | 24.9 | 28.9 | 31.6 | 32.9 | 35.3 |
| Tunisia | 5.2 | 6.4 | 8.2 | 9.7 | 10.4 | 10.8 | 11.4 |
| **Total** | **74.6** | **93.0** | **120.8** | **145.1** | **164.0** | **172.6** | **189.3** |

*Source:* IEA, Indicators for $CO_2$ Emissions, data extracted on 21 December 2018.

*Table 1A.3* *Total primary energy consumption in Mediterranean countries, 1971–2016 (Mtoe)*

| Country | 1971 | 1980 | 1990 | 2000 | 2008 | 2011 | 2016 |
|---|---|---|---|---|---|---|---|
| Albania | 1.72 | 3.07 | 2.67 | 1.79 | 2.09 | 2.22 | 2.25 |
| Algeria | 3.46 | 11.21 | 22.19 | 27.00 | 37.28 | 41.82 | 53.75 |
| Bosnia-Herz. | – | – | 7.02 | 4.35 | 5.96 | 7.15 | 6.75 |
| Croatia | – | – | 9.47 | 8.39 | 9.83 | 9.16 | 8.47 |
| Cyprus | 0.59 | 0.86 | 1.37 | 2.14 | 2.58 | 2.37 | 2.15 |
| Egypt | 7.8 | 15.1 | 32.25 | 40.11 | 72.14 | 76.23 | 86.17 |
| North Macedonia | – | – | 2.48 | 2.67 | 3.04 | 3.14 | 2.66 |
| France | 158.57 | 191.77 | 223.84 | 251.74 | 266.82 | 254.16 | 244.26 |
| Greece | 8.69 | 14.98 | 21.44 | 27.09 | 30.42 | 26.74 | 22.67 |
| Israel | 5.74 | 7.82 | 11.47 | 18.23 | 22.88 | 23.15 | 22.94 |
| Italy | 105.4 | 130.84 | 146.57 | 171.54 | 181.65 | 167.97 | 150.98 |
| Jordan | 0.49 | 1.52 | 3.27 | 4.87 | 7.07 | 7.07 | 8.98 |
| Lebanon | 1.85 | 2.47 | 1.95 | 4.91 | 5.43 | 6.35 | 7.78 |
| Libya | 1.57 | 7.05 | 11.17 | 15.82 | 17.48 | 13.58 | 15.07 |
| Malta | 0.21 | 0.32 | 0.69 | 0.68 | 0.88 | 0.83 | 0.6 |
| Montenegro | – | – | – | – | 1.21 | 1.12 | 0.97 |
| Morocco | 2.95 | 5.41 | 7.62 | 11.02 | 16.16 | 18.41 | 19.5 |
| Portugal | 6.28 | 9.99 | 16.78 | 24.59 | 24.68 | 22.85 | 22.12 |
| Serbia | – | – | 19.72 | 13.73 | 16.83 | 16.19 | 15.28 |
| Slovenia | – | – | 5.71 | 6.41 | 7.76 | 7.33 | 6.79 |
| Spain | 42.61 | 67.69 | 90.07 | 121.86 | 139.06 | 125.76 | 119.85 |
| Syria | 2.38 | 4.47 | 10.47 | 15.44 | 23.1 | 19.86 | 9.94 |
| Tunisia | 1.66 | 3.27 | 4.95 | 7.31 | 9.43 | 9.82 | 11.00 |
| Turkey | 19.54 | 31.45 | 51.44 | 76.29 | 98.42 | 112.51 | 136.72 |
| Yugoslavia | 21.91 | 33.71 | – | – | – | – | – |
| **Mediterranean** | **393.42** | **543** | **704.61** | **857.98** | **1002.2** | **975.79** | **977.65** |

*Source:* Our calculation on IEA, World Indicators, data extracted on 15 July 2019.

*Table 1A.4* *Per capita energy consumption in Mediterranean countries, 1971–2016 (toe)*

| | 1971 | 1980 | 1990 | 2000 | 2008 | 2016 |
|---|---|---|---|---|---|---|
| | | | **European Union** | | | |
| Cyprus | 1.0 | 1.7 | 2.4 | 3.1 | 3.3 | 2.5 |
| Croatia | – | – | 2.0 | 1.9 | 2.2 | 2.0 |
| France | 3.0 | 3.5 | 3.8 | 4.1 | 4.1 | 3.7 |
| Greece | 1.0 | 1.5 | 2.1 | 2.5 | 2.7 | 2.1 |
| Italy | 1.9 | 2.3 | 2.6 | 3.0 | 3.1 | 2.5 |
| Malta | 0.7 | 1.0 | 2.0 | 1.7 | 2.2 | 1.4 |
| Portugal | 0.7 | 1.0 | 1.7 | 2.4 | 2.3 | 2.1 |
| Slovenia | – | – | 2.9 | 3.2 | 3.8 | 3.3 |
| Spain | 1.2 | 1.8 | 2.3 | 3.0 | 3.0 | 2.6 |
| | | | **Europe (Candidate)** | | | |
| Albania | 0.8 | 1.1 | 0.8 | 0.6 | 0.7 | 0.8 |
| Bosnia-Herz. | – | – | 1.6 | 1.2 | 1.6 | 1.9 |
| Macedonia | – | – | 1.2 | 1.3 | 1.5 | 1.3 |
| Montenegro | – | – | – | – | 2.0 | 1.6 |
| Serbia | – | – | 2.0 | 1.7 | 2.3 | 2.2 |
| Turkey | 0.5 | 0.7 | 0.9 | 1.2 | 1.4 | 1.7 |
| | | | **North Africa** | | | |
| Algeria | 0.2 | 0.6 | 0.9 | 0.9 | 1.1 | 1.3 |
| Egypt | 0.2 | 0.3 | 0.6 | 0.6 | 0.9 | 0.9 |
| Libya | 0.7 | 2.2 | 2.5 | 3.0 | 2.9 | 2.4 |
| Morocco | 0.2 | 0.3 | 0.3 | 0.4 | 0.5 | 0.6 |
| Tunisia | 0.3 | 0.5 | 0.6 | 0.8 | 0.9 | 1.0 |
| **Middle East** | | | | | | |
| Israel | 1.9 | 2.0 | 2.5 | 2.9 | 3.1 | 2.7 |
| Jordan | 0.3 | 0.6 | 0.9 | 1.0 | 1.1 | 0.9 |
| Lebanon | 0.8 | 0.9 | 0.7 | 1.5 | 1.3 | 1.3 |
| Syria | 0.4 | 0.5 | 0.8 | 0.9 | 1.1 | 0.5 |

*Source:* IEA, World Indicators, data extracted on 2 January 2019.

*Table 2A.1 Total energy production in Mediterranean countries, 1971–2016 (ktoe)*

| Country | 1971 | 1980 | 1990 | 2000 | 2008 | 2016 |
|---|---|---|---|---|---|---|
| Albania | 2434 | 3448 | 2460 | 986 | 1148 | 1961 |
| Algeria | 41507 | 65739 | 100113 | 142229 | 162048 | 153277 |
| Bosnia-Herz. | – | – | 4604 | 3076 | 4233 | 4742 |
| Croatia | – | – | 5713 | 4263 | 4800 | 4422 |
| Cyprus | 9 | 6 | 6 | 44 | 81 | 128 |
| Egypt | 16356 | 33479 | 54871 | 53095 | 89154 | 67615 |
| Macedonia | – | – | 1257 | 1535 | 1646 | 1114 |
| France | 47610 | 52600 | 111886 | 130638 | 136761 | 131560 |
| Greece | 2085 | 3696 | 9199 | 9986 | 9858 | 6707 |
| Israel | 5938 | 154 | 424 | 642 | 3910 | 8274 |
| Italy | 19533 | 19898 | 25317 | 28173 | 32912 | 33770 |
| Jordan | 1 | 1 | 162 | 286 | 276 | 355 |
| Lebanon | 168 | 178 | 143 | 173 | 186 | 176 |
| Libya | 137513 | 96550 | 73174 | 75956 | 107245 | 29106 |
| Malta | 0 | 0 | 0 | 0 | 1 | 18 |
| Montenegro | – | – | – | – | 669 | 660 |
| Morocco | 1144 | 1415 | 1449 | 1352 | 1887 | 1783 |
| Portugal | 1385 | 1481 | 3393 | 3846 | 4473 | 6004 |
| Serbia | – | – | 13768 | 11875 | 10750 | 10696 |
| Slovenia | – | – | 3068 | 3097 | 3672 | 3585 |
| Spain | 10449 | 15773 | 34589 | 31558 | 30347 | 34125 |
| Syria | 5315 | 9501 | 22319 | 32694 | 24089 | 4222 |
| Tunisia | 4653 | 6673 | 5728 | 6635 | 7553 | 6044 |
| Turkey | 13810 | 17139 | 24833 | 26401 | 28686 | 36102 |
| Yugoslavia | 14373 | 18824 | – | – | – | – |
| **Total** | **324285** | **346554** | **498478** | **568539** | **666384** | **546446** |

*Source:* IEA, World Energy Balances, data extracted on 19 April 2019.

*Table 2A.2 Crude oil production in Mediterranean countries, 1971–2016 (ktoe)*

| Country | 1971 | 1980 | 1990 | 2000 | 2008 | 2016 |
|---|---|---|---|---|---|---|
| Albania | 1657 | 2000 | 1162 | 314 | 578 | 1056 |
| Algeria | 39069 | 54223 | 61237 | 72318 | 88187 | 72681 |
| Bosnia-Herz. | – | – | 0 | 0 | 0 | 0 |
| Croatia | – | – | 2775 | 1350 | 889 | 761 |
| Cyprus | 0 | 0 | 0 | 0 | 0 | 0 |
| Egypt | 15201 | 30257 | 46227 | 36108 | 35419 | 34711 |
| Macedonia | – | – | 0 | 0 | 0 | 0 |
| France | 2499 | 2256 | 3471 | 1811 | 1280 | 939 |
| Greece | 0 | 0 | 837 | 256 | 57 | 160 |
| Israel | 5830 | 20 | 13 | 4 | 6 | 117 |
| Italy | 1254 | 1734 | 4468 | 4692 | 5719 | 4029 |
| Jordan | 0 | 0 | 0 | 2 | 2 | 0 |
| Lebanon | 0 | 0 | 0 | 0 | 0 | 0 |
| Libya | 136112 | 92202 | 67985 | 71013 | 93099 | 20865 |
| Malta | 0 | 0 | 0 | 0 | 0 | 0 |
| Montenegro | – | – | – | – | 0 | 0 |
| Morocco | 21 | 13 | 14 | 12 | 8 | 5 |
| Portugal | 0 | 0 | 0 | 0 | 0 | 6 |
| Serbia | – | – | 1085 | 999 | 676 | 1030 |
| Slovenia | – | – | 3 | 1 | 0 | 0 |
| Spain | 127 | 1790 | 1168 | 231 | 129 | 144 |
| Syria | 5310 | 9235 | 20712 | 27791 | 19055 | 1097 |
| Tunisia | 4232 | 5816 | 4754 | 3808 | 4359 | 2507 |
| Turkey | 3529 | 2268 | 3613 | 2729 | 2134 | 2721 |
| Yugoslavia | 3023 | 4318 | – | – | – | – |
| **Total** | **217865** | **206132** | **219525** | **223439** | **251597** | **142828** |

*Source:* IEA, World Energy Balances, data extracted on 19 April 2019.

*Table 2A.3 Natural gas production in Mediterranean countries, 1971–2016 (ktoe)*

| Country | 1971 | 1980 | 1990 | 2000 | 2008 | 2016 |
|---|---|---|---|---|---|---|
| Albania | 106 | 322 | 203 | 9 | 7 | 35 |
| Algeria | 2162 | 11485 | 38854 | 69852 | 73790 | 80561 |
| Bosnia-Herz. | – | – | 0 | 0 | 0 | 0 |
| Croatia | – | – | 1619 | 1355 | 2194 | 1369 |
| Cyprus | 0 | 0 | 0 | 0 | 0 | 0 |
| Egypt | 70 | 1586 | 6733 | 14436 | 50862 | 29700 |
| Macedonia | – | – | 0 | 0 | 0 | 0 |
| France | 6051 | 6327 | 2516 | 1505 | 811 | 18 |
| Greece | 0 | 0 | 138 | 42 | 15 | 10 |
| Israel | 105 | 131 | 28 | 8 | 2801 | 7582 |
| Italy | 11023 | 10263 | 14030 | 13622 | 7580 | 4738 |
| Jordan | 0 | 0 | 102 | 213 | 152 | 86 |
| Lebanon | 0 | 0 | 0 | 0 | 0 | 0 |
| Libya | 1304 | 4223 | 5064 | 4803 | 13991 | 8087 |
| Malta | 0 | 0 | 0 | 0 | 0 | 0 |
| Montenegro | – | – | – | – | 0 | 0 |
| Morocco | 43 | 56 | 43 | 38 | 44 | 61 |
| Portugal | 0 | 0 | 0 | 0 | 0 | 0 |
| Serbia | – | – | 528 | 623 | 214 | 417 |
| Slovenia | – | – | 20 | 6 | 3 | 4 |
| Spain | 1 | 0 | 1273 | 148 | 14 | 48 |
| Syria | 0 | 40 | 1369 | 4619 | 4781 | 3039 |
| Tunisia | 1 | 351 | 331 | 1886 | 1939 | 2280 |
| Turkey | 0 | 0 | 175 | 526 | 838 | 302 |
| Yugoslavia | 1000 | 1742 | – | – | – | – |
| **Total** | **21867** | **36525** | **73026** | **113692** | **160037** | **138336** |

*Source:* IEA, World Energy Balances, data extracted on 19 April 2019.

*Table 2A.4* *Coal production in Mediterranean countries, 1971–2016 (ktoe)*

| Country | 1971 | 1980 | 1990 | 2000 | 2011 | 2016 |
|---|---|---|---|---|---|---|
| Albania | 236 | 497 | 487 | 7 | 14 | 2 |
| Algeria | 241 | 2 | 0 | 0 | 0 | 0 |
| Bosnia-Herz. | – | – | 4178 | 2458 | 3634 | 3520 |
| Croatia | – | – | 101 | 0 | 0 | 0 |
| Cyprus | 0 | 0 | 0 | 0 | 0 | 0 |
| Egypt | 0 | 0 | 0 | 36 | 0 | 0 |
| Macedonia | – | – | 1215 | 1212 | 1378 | 745 |
| France | 22962 | 13376 | 8266 | 2482 | 172 | 0 |
| Greece | 1407 | 2953 | 7119 | 8222 | 8129 | 3973 |
| Israel | 0 | 0 | 21 | 27 | 30 | 40 |
| Italy | 511 | 322 | 275 | 3 | 74 | 0 |
| Jordan | 0 | 0 | 0 | 0 | 0 | 0 |
| Lebanon | 0 | 0 | 0 | 0 | 0 | 0 |
| Libya | 0 | 0 | 0 | 0 | 0 | 0 |
| Malta | 0 | 0 | 0 | 0 | 0 | 0 |
| Montenegro | – | – | – | – | 383 | 308 |
| Morocco | 298 | 419 | 295 | 17 | 0 | 0 |
| Portugal | 152 | 73 | 115 | 0 | 0 | 0 |
| Serbia | – | – | 10172 | 8351 | 8224 | 7201 |
| Slovenia | – | – | 1350 | 1062 | 1185 | 942 |
| Spain | 6915 | 9824 | 11745 | 7966 | 4194 | 736 |
| Syria | 0 | 0 | 0 | 0 | 0 | 0 |
| Tunisia | 0 | 0 | 0 | 0 | 0 | 0 |
| Turkey | 4220 | 6153 | 11387 | 13023 | 16373 | 15498 |
| Yugoslavia | 7334 | 9624 | – | – | – | – |
| **Total** | **44276** | **43242** | **56726** | **44868** | **43790** | **32965** |

*Source:* IEA, World Energy Balances, data extracted on 19 April 2019.

*Table 2A.5 Renewable energy production in Mediterranean countries, 1971–2016 (ktoe)*

| Country | 1971 | 1980 | 1990 | 2000 | 2008 | 2016 |
|---|---|---|---|---|---|---|
| Albania | 435 | 629 | 608 | 656 | 548 | 868 |
| Algeria | 35 | 30 | 22 | 58 | 71 | 35 |
| Bosnia-Herz. | – | – | 426 | 618 | 599 | 1221 |
| Croatia | – | – | 1218 | 1557 | 1708 | 2282 |
| Cyprus | 9 | 6 | 6 | 44 | 75 | 124 |
| Egypt | 1084 | 1637 | 1911 | 2515 | 2873 | 3205 |
| Macedonia | – | – | 42 | 322 | 269 | 369 |
| France | 13666 | 14681 | 15224 | 15737 | 18680 | 23896 |
| Greece | 678 | 743 | 1105 | 1403 | 1653 | 2504 |
| Israel | 4 | 3 | 361 | 603 | 1073 | 536 |
| Italy | 5868 | 7005 | 6381 | 9597 | 18787 | 23821 |
| Jordan | 1 | 1 | 60 | 71 | 121 | 269 |
| Lebanon | 168 | 178 | 143 | 173 | 186 | 176 |
| Libya | 97 | 125 | 125 | 140 | 155 | 154 |
| Malta | 0 | 0 | 0 | 0 | 1 | 18 |
| Montenegro | – | – | – | – | 286 | 353 |
| Morocco | 781 | 927 | 1097 | 1284 | 1765 | 1645 |
| Portugal | 1233 | 1408 | 3277 | 3759 | 4340 | 5823 |
| Serbia | – | – | 1983 | 1901 | 1635 | 2049 |
| Slovenia | – | – | 491 | 788 | 835 | 1105 |
| Spain | 2748 | 2807 | 6202 | 6815 | 10316 | 17685 |
| Syria | 5 | 227 | 238 | 283 | 253 | 86 |
| Tunisia | 420 | 505 | 642 | 941 | 1255 | 1181 |
| Turkey | 6061 | 8718 | 9658 | 10102 | 9312 | 17134 |
| Yugoslavia | 3016 | 3140 | – | – | – | – |
| **Total** | **36310** | **42768** | **51222** | **59367** | **76796** | **106535** |

*Source:* IEA, World Energy Balances, data extracted on 19 April 2019.

*Table 5A.1* *Total $CO_2$ emissions in Mediterranean countries, 1971–2016 (Mt $CO_2$)*

| Country | 1971 | 1980 | 1990 | 2000 | 2008 | 2016 |
|---|---|---|---|---|---|---|
| Albania | 3.9 | 6.8 | 5.7 | 3.1 | 3.7 | 3.7 |
| Algeria | 8.6 | 27.7 | 51.2 | 61.5 | 88.1 | 127.6 |
| Bosnia-Herz. | – | – | 24.0 | 13.7 | 20.2 | 22.0 |
| Croatia | – | – | 20.3 | 16.8 | 20.2 | 15.9 |
| Cyprus | 1.7 | 2.6 | 3.9 | 6.3 | 7.6 | 6.3 |
| Egypt | 20.1 | 40.8 | 77.9 | 99.7 | 170.9 | 204.8 |
| Macedonia | – | – | 8.6 | 8.5 | 9.1 | 6.9 |
| France | 423.4 | 455.2 | 345.6 | 364.7 | 350.1 | 292.9 |
| Greece | 25.1 | 45.2 | 69.9 | 87.9 | 94.4 | 63.1 |
| Israel | 13.8 | 18.9 | 32.8 | 54.8 | 64.5 | 63.7 |
| Italy | 289.4 | 355.4 | 389.4 | 420.4 | 428.9 | 325.7 |
| Jordan | 1.4 | 4.3 | 9.2 | 14.2 | 18.5 | 23.9 |
| Lebanon | 4.6 | 6.7 | 5.5 | 14.0 | 15.8 | 23.2 |
| Libya | 3.7 | 17.6 | 25.8 | 36.8 | 42.0 | 43.3 |
| Malta | 0.7 | 1.0 | 2.3 | 2.1 | 2.7 | 1.4 |
| Montenegro | – | – | – | – | 2.7 | 2.1 |
| Morocco | 6.6 | 13.7 | 19.7 | 29.6 | 43.2 | 55.3 |
| Portugal | 14.4 | 23.8 | 37.9 | 57.9 | 52.8 | 47.4 |
| Serbia | – | – | 61.9 | 42.9 | 48.3 | 45.5 |
| Slovenia | – | – | 13.5 | 14.1 | 16.8 | 13.6 |
| Spain | 119.1 | 186.3 | 202.6 | 278.6 | 309.8 | 238.6 |
| Syria | 5.4 | 12.3 | 27.2 | 37.0 | 61.4 | 26.1 |
| Tunisia | 3.7 | 7.9 | 12.2 | 17.6 | 21.1 | 25.2 |
| Turkey | 41.7 | 71.5 | 128.8 | 201.2 | 262.4 | 338.8 |
| Yugoslavia | 61.8 | 84.2 | – | – | – | – |
| **Mediterranean** | **1049.1** | **1381.9** | **1575.9** | **1883.4** | **2155.2** | **2017.0** |

*Source:* IEA, Indicators for $CO_2$ Emissions, data extracted on 21 December 2018.

*Table 5A.2 Per capita $CO_2$ emissions in Mediterranean countries, 1971–2016 ($TCO_2$)*

| | 1971 | 1980 | 1990 | 2000 | 2008 | 2016 |
|---|---|---|---|---|---|---|
| | | **Europe** | | | | |
| Cyprus | 2.8 | 5.1 | 6.8 | 9.1 | 9.8 | 7.4 |
| Croatia | – | – | 4.3 | 3.8 | 4.5 | 3.8 |
| France | 8.1 | 8.3 | 5.9 | 6.0 | 5.4 | 4.4 |
| Greece | 2.8 | 4.6 | 6.8 | 8.1 | 8.5 | 5.9 |
| Italy | 5.4 | 6.3 | 6.9 | 7.4 | 7.2 | 5.4 |
| Malta | 2.2 | 3.1 | 6.5 | 5.5 | 6.6 | 3.1 |
| Portugal | 1.6 | 2.4 | 3.8 | 5.6 | 5.0 | 4.6 |
| Slovenia | – | – | 6.8 | 7.1 | 8.3 | 6.6 |
| Spain | 3.4 | 4.9 | 5.2 | 6.9 | 6.7 | 5.1 |
| | | **Europe (Candidate)** | | | | |
| Albania | 1.8 | 2.5 | 1.7 | 1.0 | 1.3 | 1.3 |
| Bosnia-Herz. | – | – | 5.4 | 3.6 | 5.4 | 6.2 |
| Macedonia | – | – | 4.3 | 4.2 | 4.4 | 3.3 |
| Montenegro | – | – | – | – | 4.4 | 3.4 |
| Serbia | | | 6.2 | 5.3 | 6.6 | 6.4 |
| Turkey | 1.2 | 1.6 | 2.3 | 3.1 | 3.7 | 4.3 |
| | | **North Africa** | | | | |
| Algeria | 0.6 | 1.4 | 2.0 | 2.0 | 2.5 | 3.1 |
| Egypt | 0.6 | 0.9 | 1.4 | 1.4 | 2.1 | 2.1 |
| Libya | 1.7 | 5.5 | 5.8 | 6.9 | 6.9 | 6.9 |
| Morocco | 0.4 | 0.7 | 0.8 | 1.0 | 1.4 | 1.6 |
| Tunisia | 0.7 | 1.2 | 1.5 | 1.8 | 2.0 | 2.2 |
| | | **Middle East** | | | | |
| Israel | 4.5 | 4.9 | 7.0 | 8.7 | 8.8 | 7.5 |
| Jordan | 0.7 | 1.8 | 2.6 | 2.8 | 2.9 | 2.5 |
| Lebanon | 1.9 | 2.6 | 2.0 | 4.3 | 3.8 | 3.9 |
| Syria | 0.8 | 1.4 | 2.2 | 2.3 | 3.0 | 1.4 |

*Source:* IEA, Indicators for $CO_2$ Emissions, data extracted on 21 December 2018.

*Table 5A.3 Net imports in Mediterranean countries, 1971–2016 (Mtoe)*

| Country | 1971 | 1980 | 1990 | 2000 | 2010 | 2016 |
|---|---|---|---|---|---|---|
| Albania | −0.72 | −0.38 | 0.17 | 0.83 | 0.61 | 0.45 |
| Algeria | −36.27 | −51.27 | −77.34 | −115 | −109.42 | −98.87 |
| Bosnia-Herz. | – | – | 2.44 | 1.26 | 2.02 | 2.13 |
| Croatia | – | – | 3.87 | 4.11 | 4.45 | 4.19 |
| Cyprus | 0.63 | 0.97 | 1.64 | 2.56 | 2.92 | 2.6 |
| Egypt | −8.58 | −17.16 | −20.76 | −9.67 | −11.57 | 19.32 |
| Macedonia | – | – | 1.21 | 1.1 | 1.27 | 1.57 |
| France | 117.81 | 149.01 | 119.2 | 132.49 | 132.34 | 118.25 |
| Greece | 7.69 | 13.65 | 15.32 | 21.78 | 21.3 | 18.5 |
| Israel | 0.4 | 8.46 | 11.4 | 18.17 | 20.49 | 15.09 |
| Italy | 96.38 | 116.81 | 127.27 | 152.44 | 148.48 | 121.26 |
| Jordan | 0.62 | 1.78 | 3.51 | 4.77 | 7.33 | 8.99 |
| Lebanon | 2.01 | 2.49 | 1.86 | 4.88 | 6.43 | 7.88 |
| Libya | −136.14 | −88.96 | −61.61 | −59.42 | −81.69 | −14.46 |
| Malta | 0.33 | 0.42 | 0.8 | 1.45 | 2.36 | 2.5 |
| Montenegro | – | – | – | – | 0.3 | 0.34 |
| Morocco | 2.17 | 3.96 | 6.5 | 9.93 | 16.39 | 18.59 |
| Portugal | 5.72 | 9.94 | 14.91 | 22.06 | 18.68 | 17.76 |
| Serbia | – | – | 6.09 | 1.88 | 5.2 | 4.58 |
| Slovenia | – | – | 2.62 | 3.38 | 3.58 | 3.35 |
| Spain | 35.47 | 55.33 | 60.38 | 100.19 | 106.85 | 94.5 |
| Syria | −2.54 | −4.06 | −11.14 | −17.81 | −4.37 | 5.89 |
| Tunisia | −2.56 | −2.49 | −0.87 | 0.81 | 2.06 | 5.2 |
| Turkey | 6.11 | 14.38 | 27.78 | 50.66 | 75.92 | 105.67 |
| Yugoslavia | 6.94 | 15.23 | – | – | – | – |

*Source:* IEA, World Indicators, data extracted on 15 July 2019.

# References


Abu-Dayyeh, Ayoub (2018), 'Jordan: Overcoming energy insecurity', in David Ramin Jalivand and Kirsten Westphal (eds), *The Political and Economic Challenges of Energy in the Middle East and North Africa*, New York, NY: Routledge.

Agovino, M., S. Bartoletto and A. Garofalo (2019), 'Modelling the relationship between energy intensity and GDP for European countries: An historical perspective (1800–2000)', *Energy Economics*, 82, 114–34.

Alabi, L.O. (2020), 'Oil price spikes amid US–Iran tensions', *Financial Times*, 9 January, accessed 14 January 2020 at https://www.ft.com/content/1155034d-6d5b-4388-bfa3-0c51760a01be.

Alfieri, M. (2019), 'Mattei l'apripista', Eniday Archives, accessed 4 February 2019 at https://www.eniday.com/it/human_it/enrico-mattei-industria-petrolifera-iran.

Allen, Robert (2009), *The British Industrial Revolution in Global Perspective*, Cambridge: Cambridge University Press.

Anderson, Irvine (1981), *Aramco, the United States, and Saudi Arabia*, Princeton, NJ: Princeton University Press.

Ansani, Andrea and Daniele, Vittorio (2014), 'Le economie del Mediterraneo: Lo sviluppo economico e le disuguaglianze', in Adalgiso Amendola and Eugenia Ferragina (eds), *Economia e istituzioni dei paesi del Mediterraneo*, Bologna: Il Mulino.

Arab Gas Pipeline (AGP) (n.d.), Arab Gas Pipeline (AGP), Jordan, Syria, Lebanon, *Hydrocarbons Technology*, accessed 25 February 2020 at https://www.hydrocarbons-technology.com/projects/arab-gas-pipeline-agp/.

Arezki, R. and Blanchard, O.J. (2014), 'The 2014 oil price slump: Seven key questions', IMFdirect – The IMF Blog, 22 December.

Atallah, Sami and Bassam Fattouh (2019), 'Introduction', in Sami Atallah and Bassam Fattouh (eds), *The Future of Petroleum in Lebanon: Energy, Politics and Economic Growth*, London, UK and New York, NY, USA: I.B. Tauris.

Bahgat, G. (2005), 'Energy partnership: Israel and the Persian Gulf', *Energy Policy*, 33, 671–7.

Balta, Paul (2000), *Méditerranée. Défis et enjeux*, Paris: L'Harmattan.

Barcelona Declaration adopted at the Euro-Mediterranean Conference, 27–28 November 1995, accessed at https://ec.europa.eu/research/iscp/pdf/policy/barcelona_declaration.pdf.

Bartoletto, Silvana (2004), 'Dalla legna al carbon fossile: I consumi di combustibile a Napoli nel corso dell'Ottocento', *Mélanges de l'École française de Rome. Italie et Méditerranée*, 116(2), 705–21.

Bartoletto, Silvana (2005), 'I combustibili fossili in Italia dal 1870 ad oggi', *Storia Economica*, 2, 281–327.

Bartoletto, Silvana (2007), 'L'energia delle città. Percorsi di ricerca, muovendo dal caso di Napoli', in Gabriella Corona and Simone Neri Serneri (eds), *Storia e ambiente. Città, risorse e territori nell'Italia contemporanea*, Rome: Carocci, pp. 218–33.

Bartoletto, Silvana (2012a), 'Patterns of energy transitions: The long-term role of energy in the economic growth of Europe', in Nina Möllers and Karin Zachmann (eds), *Past*

*and Present Energy Societies: How Energy Connects Politics, Technologies and Cultures*, Bielefeld: Transcript Verlag.
Bartoletto, Silvana (2012b), 'La dipendenza energetica nei paesi del Mediterraneo', in Paolo Malanima (ed.), *Rapporto sulle Economie del Mediterraneo. Edizione 2012*, Bologna: Il Mulino.
Bartoletto, Silvana (2013a), 'Energy and economic growth in Europe: The last two centuries', in Bruno Chiarini and Paolo Malanima (eds), *From Malthus Stagnation to Sustained Growth: Social, Demographic and Economic factors*, Basingstoke: Palgrave Macmillan, pp. 52–70.
Bartoletto, Silvana (2013b), 'L'energia: I consumi finali nei paesi del Mediterraneo', in Paolo Malanima (ed.), *Rapporto sulle Economie del Mediterraneo. Edizione 2013*, Bologna: Il Mulino.
Bartoletto, Silvana (2016a), *Energia e crescita economica nei paesi del Mediterraneo*, Milan: Bruno Mondadori.
Bartoletto, Silvana (2016b), 'Economic growth and oil routes in the Mediterranean countries', in Alain Beltran (ed.), *Les Routes du Pétrole* (Oil Routes), Brussels: P.I.E. Peter Lang.
Bartoletto, Silvana and Malanima, Paolo (2014), 'L'energia nei paesi del Mediterraneo 1950–2010', in Amendola Adalgiso and Eugenia Ferragina (eds), *Economia e Istituzioni dei Paesi del Mediterraneo*, Bologna: Il Mulino.
Bartoletto, S. and Rubio-Varas, M.d.M. (2008), 'Energy transition and $CO_2$ emissions in Southern Europe: Italy and Spain (1861–2000)', *Global Environment. Journal of History and Natural and Social Sciences*, 2, 46–81.
Baumeister, C. and L. Kilian (2016), 'Forty years of oil price fluctuations: Why the price of oil may still surprise us', *Journal of Economic Perspectives*, 30(1), 139–60.
Bellodi, Leonardo (2019), 'Chi (non) controlla il petrolio (non) controlla la Libia', in *Dalle Libie all'Algeria affari nostri, Limes*, 6.
Bellù, L.G. and Liberati, P. (2006), *Inequality Analysis: The Gini Index*, Food and Agriculture Organization of the United Nations, FAO.
Benoit, Guillaume and Comeau, Aline (2005), *A Sustainable Future for the Mediterranean: The Blue Plan's Environment and Development Outlook*, Abingdon, UK and New York, NY, USA: Earthscan.
Bloomberg New Energy Finance (2018), *Global Trends in Renewable Energy Investment 2018*, Frankfurt School-UNEP Centre/BNEF.
Bloomfield, J., Copsey, N. and Rowe, C. (2011), 'Renewable energy in the Mediterranean', European Union, accessed at https://cor.europa.eu/en/engage/studies/Documents/renewable-energy-mediterranean.pdf.
Blue Plan (2001), 'Urbanisation in the Mediterranean region from 1950 to 1995', *Blue Plan Papers*, no. 1, October.
Blue Plan (2012), 'Towards a Euro-Mediterranean sustainable urban strategy (EMSUS) within the framework of the Union for the Mediterranean: A diagnosis of the Mediterranean cities situation', Contribution to the urban working group of the Secretariat of the Union for the Mediterranean, January.
Bongiorni, R. (2013), 'Libia sull'orlo del caos: Crolla l'export di petrolio, major in fuga. L'Eni intenzionata a restare', *Il Sole 24 ore*, 31 October.
Campbell, Robert W. (1968), *The Economics of Soviet Oil and Gas*, Baltimore, MD: Johns Hopkins University Press.
Capasso, Salvatore (2018), 'I paesi del Mediterraneo tra disuguaglianza e convergenza: Sviluppo economico, sostenibilità e migrazioni', in Eugenia Ferragina (ed.), *Rapporto sulle Economie del Mediterraneo. Edizione 2018*, Bologna: Il Mulino.

Chancel, L. and Piketty, T. (2015), 'Carbon and inequality: From Kyoto to Paris. Trends in the global inequality of carbon emissions (1998–2013) & prospects for an equitable adaptation fund', WID. world Working paper series, no. 7, November.
Chauhan, V. and Sen, A. (2018), 'Saggezza non convenzionale', *Energia*, 3, 12–14.
Clô, Alberto (2016), 'La sicurezza energetica: Vero o falso problema', in Massimo Nicolazzi and Nicolò Rossetto (eds), *L'età dell'abbondanza: Come cambia la sicurezza energetica*, Milan: ISPI.
Clô, Alberto (2017), *Energia e Clima: L'altra Faccia della Medaglia*, Bologna: Il Mulino.
Clô, A. (2019), 'Bilancio di fine anno', *Energia*, accessed 28 December 2019 at www.rivistaenergia.it.
Clô, Alberto (2000), *Economia e Politica del Petrolio*, Bologna: Compositori.
Colitti, Marcello (1979), *Energia e Sviluppo in Italia: La Vicenda di Enrico Mattei*, Bari: De Donato.
Cuñado, Juncaland and Fernando Pérez de Gracia (2003), 'Do oil price shocks matter? Evidence for some European countries', *Energy Economics*, 25, 137–54.
Ellinas, Charles (2018), 'The Eastern Mediterranean: An energy region in the making', in David Ramin Jalivand and Kirsten Westphal (eds), *The Political and Economic Challenges of Energy in the Middle East and North Africa*, New York, NY: Routledge.
ENEA (2001), 'I consumi energetici di biomasse nel settore residenziale in Italia nel 1999', Rome: ENEA.
Engemann, Kristie M., Kevin L. Kliesen and Michael T. Owyang (2011), 'Do oil shocks drive business cycles? Some U.S. and international evidence', *Macroeconomic Dynamics*, 15, 498–517.
European Commission (2018), Renewable Energy – Recast to 2030 (RED II), accessed 25 February 2020 at https://ec.europa.eu/jrc/en/jec/renewable-energy-recast-2030-red-ii.2
European Commission (2019), The European Green Deal, Brussels, 11 December.
European Commission (n.d.), Renewable Energy Directive, accessed 25 February 2020 at https://ec.europa.eu/energy/en/topics/renewable-energy/renewable-energy-directive/overview.
European Parliament (2009), Directive 2009/28/EC of the European Parliament and of the Council of 23 April 2009 on the promotion of the use of energy from renewable sources and amending and subsequently repealing Directives 2001/77/EC and 2003/30/EC.
European Parliament (2014), 'Shale gas and EU energy security', Briefing, December.
European Union (2018), Directive 2018/2001 of the European Parliament and of the Council of 11 December 2018 on the promotion of the use of energy from renewable sources.
European Union (2019), 'Shedding light on energy in EU. A guided tour of energy statistics', May.
European Union (n.d.), Use of international credits, accessed 24 February 2020 at https://ec.europa.eu/clima/policies/ets/credits_en.
Evans, Joanne, Filippini, Massimo and Hunt, Lester (2013), 'The contribution of energy efficiency towards meeting $CO_2$ targets', in Roger Fouquet (ed.), *Handbook on Energy and Climate Change*, Cheltenham, UK and Northampton, MA, USA: Edward Elgar Publishing.

Fattouh, Bassam and Economou, Andreas (2019), ‘Demand shocks, supply shocks and oil prices: Implications for OPEC’, Oxford: Oxford Institute for Energy Studies, 26 June.

Fattouh, Bassam and El-Katiri, Laura (2019), ‘Lebanon’s gas trading options’, in Sami Atallah and Bassam Fattouh (eds), *The Future of Petroleum in Lebanon. Energy, Politics and Economic Growth*, London, UK and New York, NY, USA: I.B. Tauris, pp. 155–71.

Ferragina, Annamaria and Nunziante, Giulia (2018), ‘Disparità e ineguaglianze tra i territori all’interno dell’area mediterranea’, in Eugenia Ferragina (ed.), *Rapporto sulle Economie del Mediterraneo. Edizione 2018*, Bologna: Il Mulino.

First, Ruth (1974), *Lybia: The Elusive Revolution*, London: Penguin.

Fouquet, Roger (2008), *Heat, Power and Light: Revolutions in Energy Services*, Cheltenham, UK and Northampton, MA, USA: Edward Elgar Publishing.

Fouquet, R. (2010), ‘The slow search for solutions: Lessons from historical energy transitions by sector and service’, *Energy Policy*, 38, 6586–96.

Fouquet, Roger (2011), ‘The sustainability of “sustainable” energy use: Historical evidence on the relationship between economic growth and renewable energy’, in I. Galarraga, M. González-Eguino and A. Markandya (eds.) *Handbook of Sustainable Energy*, Cheltenham, UK and Northampton, MA, USA: Edward Elgar Publishing.

Fouquet, Roger (2013), ‘Low-carbon economy: Dark age or golden age’, in Roger Fouquet (ed.), *Handbook on Energy and Climate Change*, Cheltenham, UK and Northampton, MA, USA: Edward Elgar Publishing.

Fouquet, R. (2017), ‘Make low-carbon energy an integral part of the knowledge economy’, *Nature*, 551, S141.

Fouquet, Roger (2019), ‘Introduction to the Handbook on Green Growth’, in Roger Fouquet (ed.), *Handbook on Green Growth*, Cheltenham, UK and Northampton, MA, USA: Edward Elgar Publishing.

Fouquet, Roger and Hippe, Ralph (2019), ‘The transition from a fossil fuels economy to a knowledge economy’, in Roger Fouquet (ed.), *Handbook on Green Growth*, Cheltenham, UK and Northampton, MA, USA: Edward Elgar Publishing.

Gilpin, Robert (1975), *US Power and the Multinational Corporation*, New York, NY: Basic Books.

Gini, C. (1921), ‘Measurement of inequality of incomes’, *Economic Journal*, 31, 124–6.

Hall, R. (2020), ‘“No, no America”: Thousands march in Iraq to demand exit of US forces’, accessed 24 January 2020 at https://www.independent.co.uk/news/world/middle-east/iraq-protests-us-troops-moqtada-al-sadr-soleimani-baghdad-a9300256.html.

Hamilton, James D. (1983), ‘Oil and the macroeconomy since World War II’, *Journal of Political Economy*, 91, 228–48.

Hamilton, James D. (2011a), ‘Nonlinearities and the macroeconomic effects of oil prices’, *Macroeconomic Dynamics*, 15, 364–78.

Hamilton, James D. (2011b), ‘Historical oil shocks’, NBER Working Papers Series, February.

Hamilton, James D. (2013), ‘Oil prices, exhaustible resources and economic growth’, in Roger Fouquet (ed.), *Handbook on Energy and Climate Change*, Cheltenham, UK and Northampton, MA, USA: Edward Elgar Publishing.

Hepburn, Cameron and Bowen, Alex (2013), ‘Prosperity with growth, economic growth, climate change and environmental limits’, in Roger Fouquet (ed.), *Handbook*

*on Energy and Climate Change*, Cheltenham, UK and Northampton, MA, USA: Edward Elgar Publishing.
IEA, Indicators for $CO_2$ Emissions, at www.iea.org.
IEA, World Energy Balances, at www.iea.org.
IEA, World Indicators, at www.iea.org.
IEA (2016), *Energy Efficiency Indicators*, Paris: OECD.
IEA (2016), *Key World Energy Statistics*, Paris: OECD.
IEA (2018), *Key World Energy Statistics*, Paris: OECD.
IEA (2018), *World Energy Balances*, Paris: OECD.
IEA (2019), *World Energy Balances*, Paris: OECD.
IMF (2013), Libya country report, May.
IMF (2015), Jordan, Country report no. 15/225, August.
IPCC (2018), 'Global warming of 1.5°C', accessed at https://www.ipcc.ch/site/assets/uploads/sites/2/2018/07/SR15_SPM_version_stand_alone_LR.pdf.
Jalilvand, David Ramin and Kirsten Westphal (2018), 'Introduction', in David Ramin Jalilvand and Kirsten Westphal (eds), *The Political and Economic Challenges of Energy in the Middle East and North Africa*, New York, NY: Routledge.
Jentleson, B.W. (1984), 'From consensus to conflict: The domestic political economy of East–West energy trade policy', *International Organization*, 38(4), 625–60.
Jiménez-Rodríguez, Rebeca and Marcelo Sanchez (2005), 'Oil price shocks and real GDP growth: Empirical evidence for some OECD countries', *Applied Economics*, 37, 201–28.
Kander, Astrid, Malanima, Paolo and Warde, Paul (2013), *Power to the People: Energy in Europe over the Last Five Centuries*, Princeton, NJ, USA and Woodstock, UK: Princeton University Press.
Kapstein, Ethan B. (1990), *The Insecure Alliance: Energy Crises and Western Politics since 1944*, Oxford: Oxford University Press.
Keramane, Abdennour (2001), 'L'energia e la sua distribuzione: Petrolio, gas naturale, elettricità', in *Il Mediterraneo: Economia e Sviluppo*, Milan: Jaca Book.
Kilian, L. and Vigfusson, R.J. (2011), 'Are the responses of the U.S. economy asymmetric in energy price increases and decreases?' , Quantitative Economics, 2, 419–53.
Kim, Dong Heon (2012), 'What is an oil shock? Panel data evidence', *Empirical Economics*, 43, 121–43.
Labbate, Silvio (2016), *Illusioni Mediterranee: Il Dialogo Euro-Arabo*, Milan: Mondadori Education.
Lomborg, Bjorn (2001), *The Skeptical Environmentalist: Measuring the Real State of the World*, New York, NY: Cambridge University Press.
Luciani, Giacomo (1976), *L'Opec nella Economia Internazionale*, Turin: Einaudi.
Luciani, Giacomo (1984), 'The Mediterranean and the energy picture', in Giacomo Luciani (ed.), *The Mediterranean Region*, London: Croom Helm.
Matus, K., Nam, K., Selin, N.E., Lamsal, L.N., Reilly, J.M. and Paltsev, S. (2012), 'Health damages from air pollution in China', *Global Environmental Change*, 22, 55–66.
MEDENER-OME (2018), Les énergies renouvelables en Mediterranée: Tendances, perspectives et bonne pratiques, MEDENER-OME, accessed at www.medener.org.
Medlock, K.B. and Soligo, R. (2001), 'Development and end-use energy demand', *Energy Journal*, 22(2), 77–105.
Michaels, L. and Tal, A. (2015), 'Convergence and conflict with the National Interest: Why Israel abandoned its climate policy', *Energy Policy*, 87, 480–85.

Ministero dello Sviluppo Economico (2018), accessed at https://dgsaie.mise.gov.it/gas_naturale_importazioni.php, several years.
Negri, A. (2019), 'Il grande gioco del gas nell'Egeo: Nuove alleanze e conflitti', *Il Quotidiano del Sud*, 4 August, accessed at https://www.lantidiplomatico.it/.
Nordhaus, W.D. (1992), 'Lethal Model 2: The limits to growth revisited', *Brooking Papers on Economic Activity*, 2, 1–59.
OECD (2018), 'The long view: Scenarios for the world economy to 2060', *OECD Economic policy paper*, July, no. 22.
O'Garra, Tanya (2013), 'Individual consumers and climate change: Searching for a new moral compass', in Roger Fouquet (ed.), *Handbook on Energy and Climate Change*, Cheltenham, UK and Northampton, MA, USA: Edward Elgar Publishing.
Oumansour, Brahim (2019), 'Chi comanda in Algeria', in Limes (ed.), *Dalle Libie all'Algeria Affari Nostri, Limes, Rivista Italiana di Geopolitica*, 6/2019.
Pace, Giuseppe (2003), 'Il contesto mediterraneo', in Giuseppe Pace (ed.), *Economie Mediterranee. Rapporto 2003*, Napoli: Edizioni Scientifiche Italiane.
Philbin, B. (2013), 'Where Syria fits in the world's oil supply system', *The Wall Street Journal*, 27 August.
Quadri, Susanna (2012), *Energia Sostenibile*, Turin: Giappichelli.
Reed, Stanley and Krauss, Clifford (2020), 'Oil prices are slow to reflect U.S.–Iran tensions', *The New York Times*, 7 January.
Rettig, Elai (2018), 'The impact of natural gas discoveries upon Israeli politics, geopolitics, and socioeconomic discourse', in David Ramin Jalivand and Kirsten Westphal (eds), *The Political and Economic Challenges of Energy in the Middle East and North Africa*, New York, NY: Routledge, pp.169–81.
Reynolds, D.B. (1994), 'Energy grades and economic growth', *The Journal of Energy and Development*, 19(2), 245–64.
Rubio-Varas, Maria del Mar and De la Torre, Joseba (2017), 'How did Spain become the major US nuclear client?', in Maria del Mar Rubio-Varas and Joseba De la Torre (eds), *The Economic History of Nuclear Energy in Spain*, Basingstoke: Palgrave Macmillan.
Scholl, Ellen (2018), 'Turkey: In between global trends and regional politics', in David Ramin Jalilvand and Kirsten Westphal (eds), *The Political and Economic Challenges of Energy in the Middle East and North Africa*, New York, NY: Routledge.
Stern, D.I. (2010), 'Energy quality', *Ecological Economics*, 69(7), 1471–8.
Stern, D.I. (2011), 'The role of energy in economic growth', *Annals of New York Academy of Sciences*, 1219, 26–51.
Talbot, V. and A. Varvelli (2019), 'Turchia sul fronte Libia con armi e gas', ISPI, 19 December, accessed at https://www.ispionline.it/.
Thomas, N. and S. Nigam (2018), 'Twentieth-century climate change over Africa: Seasonal hydroclimate trends and Sahara Desert expansion', *Journal of Climate*, 31(9), 3349–70.
Unione Petrolifera (2018), 'Relazione annuale', accessed at www.unionepetrolifera.it/download/relazione-annuale-2018.
US Energy Information Administration, Country analysis brief: various countries, accessed at accessed at www.eia.doe.gov.
US Energy Information Administration, Syria, full report, last updated 24 June 2015, accessed at www.eia.doe.gov.
US Energy Information Administration, World oil transit chokepoints, 25 July 2017.
US Energy Information Administration, *Country Analysis Executive Summary: Algeria*, 25 March 2019.

Varvelli, A. (2018), 'Italia-Libia: Un contributo vero alla stabilizzazione', in *Affari Internazionali*, 12 October, accessed at www.affarinternazionali.it.
Voigt, S., De Cian, E., Schymura, M. and Verdolini, E. (2014), 'Energy intensity developments in 40 major economies: Structural change or technology improvement?', *Energy Economics*, 41, 47–62.
Warde, Paul (2007), 'Energy consumption in England and Wales, 1560–2004', Naples: Consiglio Nazionale delle Ricerche, Istituto di Studi sulle Società del Mediterraneo.
Yergin, Daniel (1991), *The Prize: The Epic Quest for Oil, Money and Power*, New York, NY: Simon & Schuster.
Zupi, Marco (2017), 'La situazione occupazionale sulle sponde del Mediterraneo', *Osservatorio di Politica Internazionale*, no. 73, May.

# Index